SEEDS and CUTTINGS

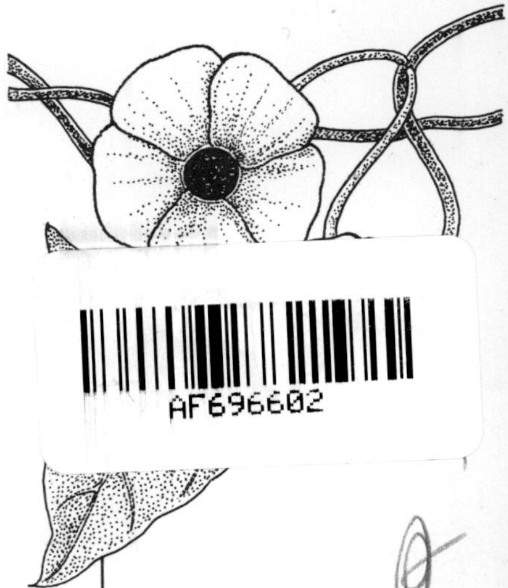

BOOKS BY H. PETER LOEWER:

Indoor Water Gardner's How-to Handbook
Bringing the Outdoors In
ILLUSTRATOR OF: *Growing Unusual Fruit*

SEEDS and CUTTINGS

written and illustrated by
H. Peter Loewer

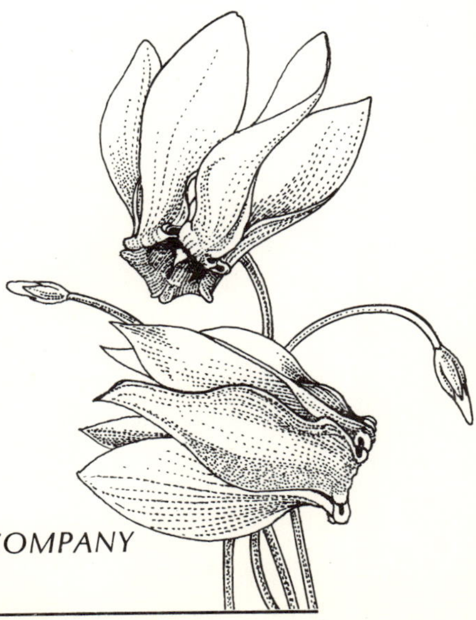

WALKER AND COMPANY
NEW YORK

First published in the United States of America in 1975 by the Walker Publishing Company, Inc.

Published simultaneously in Canada by Fitzhenry & Whiteside, Limited, Toronto.

ISBN: 0-8027-0480-8

Library of Congress Catalog Card Number: 74-31913

Printed in the United States of America.

10 9 8 7 6 5 4 3 2 1

If you can look
into the seeds of time,
and say which grain will grow
and which will not.

Macbeth, Act I

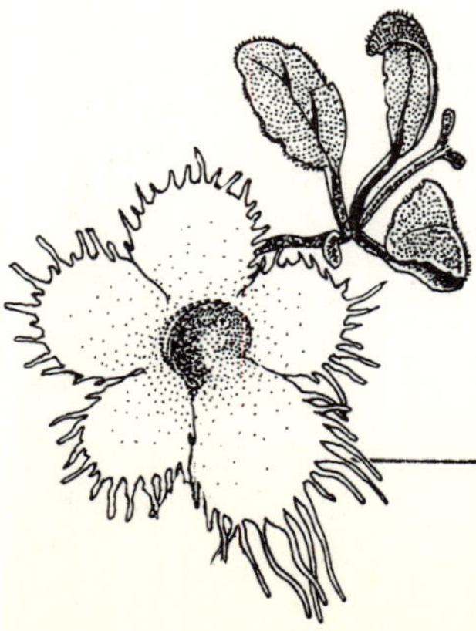

Contents

Preface

The *New York Times* of May 31, 1975, carried a lead story on its editorial page entitled: "Flower Smuggler, Drop That Pistil!" It concerned the 25-30,000 plants from all over the world, now threatened with extinction from the unsatiable demands of so-called plant collectors (and greedy land developers). They are plant pirates in every sense of the words: people who think nothing of stripping mountains of their greenery, deserts of their cacti, and the back yards of Florida homes for ornaments to grace their living rooms. For them it is an endless search for ego gratification or just plain money.

This book offers an alternate solution to increasing plant collections by pointing out the excitement of growing plants from seed, and the scientific and practical enjoyment of using the cellular tricks of plants evolution to grow new specimens from simple cuttings of leaves, stems, and roots. In addition, there is a chapter on ferns, which reproduce with a complicated process only discovered in the last century, and a chapter on cacti and bromeliads.

For day-to-day reference, I have put two extensive tables in the appendix: one lists methods of propagation for a large number of houseplants, and the other lists plants that may be started from seed at any time of the year.

I would appreciate hearing from any readers who have new experiences with plant propagation or who, perhaps, discover old methods that have been forgotten over the years. Just send all communications to me at Walker and Company, 720 Fifth Avenue, New York, N.Y. 10019.

Again heartfelt thanks go to my editor, Richard Winslow, for his unerring sense of timing in pushing the publisher's panic button, and his constant help and encouragement; and to my wife, Jean, for stepping over and around the countless heaps of rubble, soil, seed-husks, and paraphernalia that follow me from room to room.

—H. Peter Loewer

Introduction

Many types of plant reproduction are covered in this book, but they all fall within two categories: *sexual*, or reproduction from the direct union of male and female plant cells, which leads to seeds (or spores in the case of ferns); and *asexual* (also called *vegetative* reproduction), when new plants are formed from certain parts of mature plants—the stems, the roots, or the leaves.

SEEDS

A seed is a bit like a spaceship that drifts or falls through the air rather than hurtling across empty space. Both seed and ship must provide a wall of protection for life within and enough food and oxygen to support life until the ship or seed is opened. The hazards facing a space traveler on an alien planet could not be more formidable than those that a seed faces as it tries to start to grow on earth

—wind, snow, rain, and hail; the dry heat of a desert sun or the rigid cold of a northern winter; the rumbling stomach of a bird or beast; and the endless creeping and crawling enemies that exist within and on the earth's skin.

Yet a few seeds of the thousands cast manage to survive, to germinate and grow, and to produce seeds of their own, thus beginning the cycle again.

Lotus pod with seeds

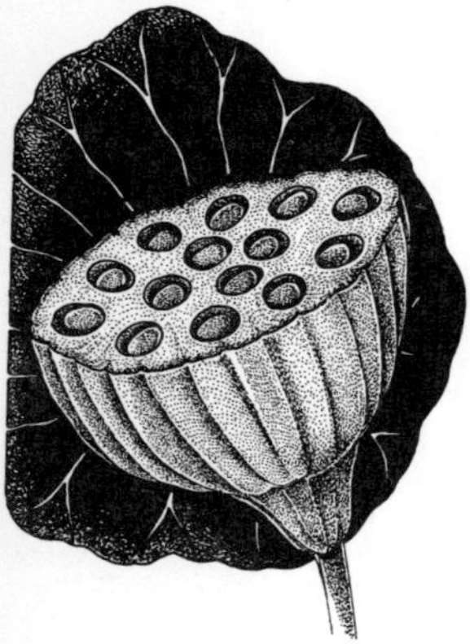

Seeds come as large as a coconut or as small as the dust motes of a begonia flower, and all shapes and sizes in between. Some die within a few days of falling to the earth, like the silver maple's seed of our northeastern forests; many survive for years waiting for just the right combination of moisture and germination. A few, like the regal lotus, may lie within a tomb for centuries, buried as an offering to eternity by a culture long departed, to spring to life in a laboratory of the twentieth century.

Seeds of wheat provide more food for the human race than any other plant or animal, with rice taking second place. We consume vast numbers of pecans, peanuts, and pistachios, all seeds. About one-fourth of the human energy demands of the United States are provided by cereal seeds; and three-quarters of human diet in the rest of the world. We mash, grind, boil, and compress seeds into margarine, medicines, cosmetics, and alcoholic beverages. And, for the gardener, seeds serve as an incomparable source of beauty in and around the home.

CUTTINGS

While seeds have been utilized by man since neolithic times, the arts of asexual propagation have a much shorter history.

Early man must have noticed how branches, when pressed on branches of other trees by natural forces, will sometimes slowly bind together permanently. This phenomena was observed by the first agricultural people and, no doubt, inspired the invention of grafting. By biblical times, grafting techniques were already producing substantial vineyards and orchards of the day. Grafting was fully documented by the Romans and has continued to develop as the basis for many modern agricultural practices.

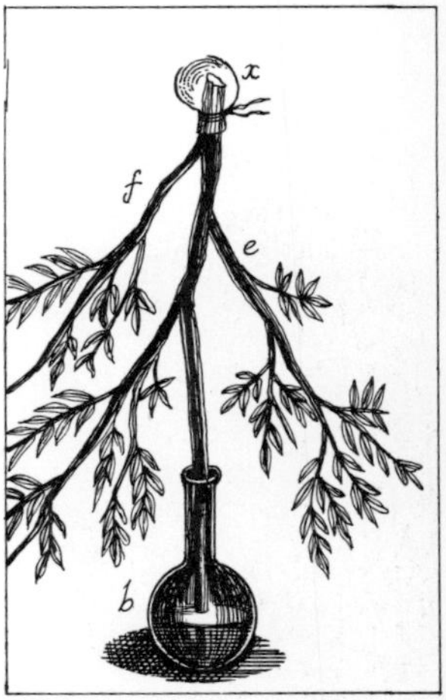

18th Century experiment showing a tree branch immersed upside down and drawing water to a bladder tied on top.

The other means of vegetative propagation were not discovered until the rise of modern science in the seventeenth century, when the hundreds of gentlemen naturalists began to record and document their avid experiments with plants. Asexual propagation, it's interesting to note, had its setbacks. Even in Darwin's time plant propagation techniques were considered perverse practices, violating the laws of nature and God in the eyes of more than one religious group.

The concept of seed formation is easy to grasp, since it follows the same patterns of most plants and animals for reproduction. Asexual reproduction, or "cloning," the biological term recently popularized by science fiction authors, becomes a bit more amazing in its implications. Since every living cell is programmed genetically to carry all the information needed for the development of a complete organism, it is theoretically possible to produce a beet or a person from a single beet cell or a single human cell. Gardeners, amateur and professional, for years have been cloning things whenever they grow plants from cuttings, and, the scientists assure us, with enough time, study, and money, someday they will be able to clone human beings. Somehow, armies of philodendrons raised from one leaf cell are not quite as frightening as vast hordes of people, all identical, produced from a single cell of one man.

CHAPTER 1

What Is a Seed?

Seed formation is an essential step in the survival cycle of most plants, and this process begins in the flower.

FLOWERS

All flowers are based on one general plan. Many have succeeded in hiding their similarities by evolutionary changes, but the basic parts are there, if we look closely enough. The following illustration shows the floral organs of a typical flower:

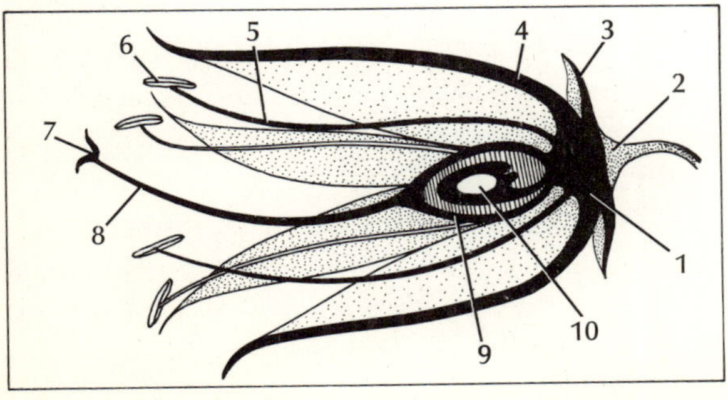

The *receptacle* (1) is attached to the plant stem by a special stalk known as the *pedicel* (2). The receptacle holds the *sepals* (3), which are collectively called the *calyx*. Occasionally sepals are brightly colored, as in tulips and lilies, but usually they are green. Sepals form a protective layer over the *petals* (4), which are usually brightly colored to form the initial attraction for insects, and are collectively called the *corolla*. Inside the corolla there are the *stamens* or male floral organs. The stamens consist of the *filament* (5) and the *anther* (6). The anther is the site of pollen formation. The female floral organ is called the *pistil* and is composed of the *stigma* (7), which receives the pollen and is held by the *style* (8), which in turn is attached to the *ovary* (9), where the *ovules* (10) or seeds develop.

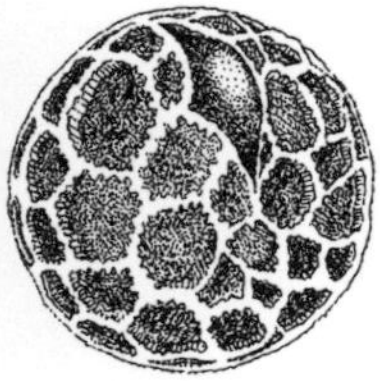

Pollen grain

Pollinators, such as the bee, are attracted by the color and scent of the flower. Their goal is the nectar, produced by the flower. The bee picks up bits of pollen that stick to the hairs on its body, mouth, and legs, and it inadvertently deposits the pollen on flower stigmas. Fertilization occurs and seeds are started.

Many flowers consist of both male and female floral organs and are called *complete* flowers. The rose is an obvious example.

A few have separate male and female flowers on the same plant, as are found in the poinsettia.

Poinsettia flowers

The bittersweet vine of North America is an example of a plant that is either entirely male or female. Every year gardeners wait in vain for the bright orange berries, because the nurseryman did not sell them two plants, one male (for the pollen) and one female (for the fruit). The hollies present a similar problem to purchasers of only one plant.

When you have flowering houseplants with complete flowers (both male and female) such as morning glories, you can do your own pollinating by dabbing the anther with a Q-Tip or a camel's hair brush, and transferring the pollen to the stigma.

Some flowers refuse to be pollinated by their own pollen, and pollen from another plant of the same species must be used. The common day lily is a prime example, and many bromeliads will not self-pollinate.

At times the final result will be slightly disappointing as many of the large and beautiful flowers of today are hybrids,* and each time you cross-pollinate, you travel back to the original flower and lose much of the impressive size and color. This is a fascinating hobby and many amateurs have come up with beautiful (and sometimes valuable) hybrids.

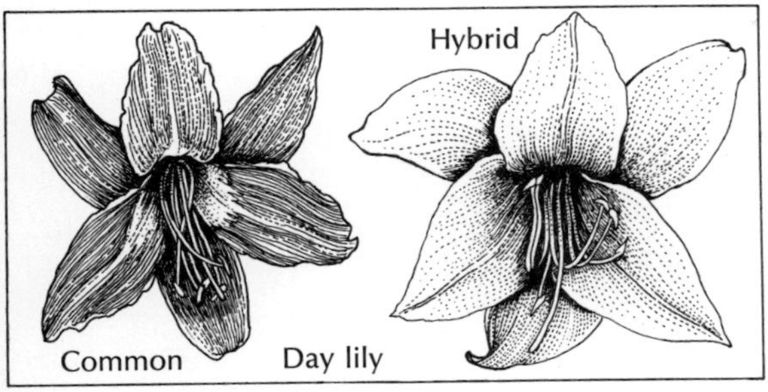

*A plant produced by parents belonging to different species or to different strains of the same species. A hybrid takes characteristics from each parent and occasionally turns out to be more valuable than either. Most of our spectacular garden plants are produced from hybridization. Unfortunately, they rarely breed true and must be propagated by asexual means.

SEEDS

Although found in a mystifying variety of shapes, sizes, and colors, all seeds consist of the embryo (1) or baby plant and the seed coat (2) or protective shell.

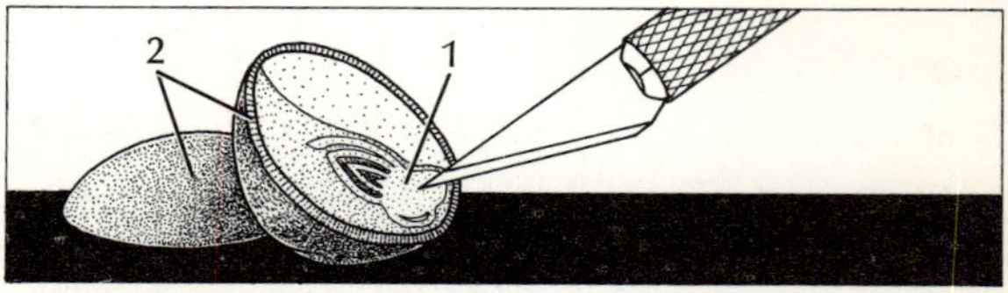

The embryo has a rudimentary stem, one or two seed leaves, or cotyledons, and a tiny root. In addition, the seed stores food in the form of starch, sugars, fats, and proteins. These foods will give to the infant plant its start in life, but at this stage, the seed is essentially a young plant in a state of arrested development.

Germination

In order for the seed to begin to grow, germination must occur and three items are essential: water, a favorable temperature, and oxygen. If any of these things are missing, the seed will not germinate. In addition, after the first three conditions are met, light will play a key role.

Water softens the hard seed coat, making it easier for the embryo to break through. It also causes the seed to swell so that it can burst out of the seed coat. Water helps to pass oxygen to the embryo for the life process to begin.

There is an optimum temperature for seed germination. For many crop plants it is 10-15°C (50-60°F), but most houseplants require 18°-27°C (65°-80°F).

Finally, many seeds are very small, and if buried too deep in the soil, the seedling would perish before reaching the surface and sunlight. These seeds will not germinate without the added stimulation of light.

Dormancy of Seeds

Many plants that are native to the temperate zone will not germinate until they have passed through a state of dormancy. This protects the plant from sprouting during a January thaw only to expire in a February blizzard. The two most common forms of dormancy are impermeability of the seed coat, which prevents water from entering, and/or a biological thermostat in the embryo. In some species, only one such mechanism exists; others may have both.

If a thick seed coat is the problem, it's easily solved by nicking the seed with a file. Many seed packets will give this advice.

If thermostat dormancy is the problem, it's solved by a process called *stratification*. The seeds have to be placed in a bed of moist sand or peatmoss and exposed to temperatures of 1-6°C (34-43°F) for several months in an effort to duplicate the temperatures of winter. The family refrigerator works very well. We'll mention this again in chapter 2. It's a good thing to remember, especially when dealing with seeds collected in the wild.

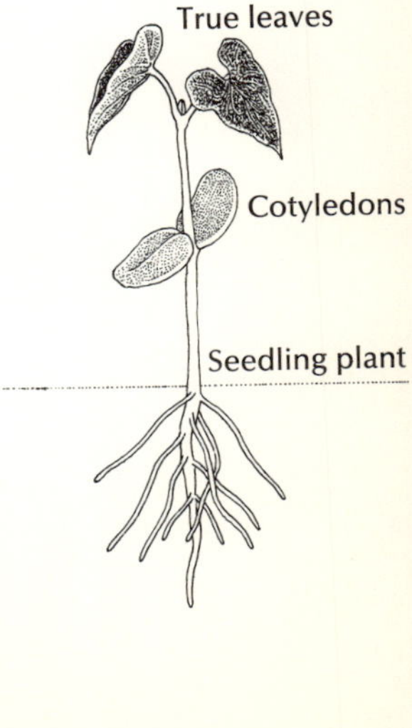

SEEDLINGS.

The young plant, a seedling, is now emerging from the soil. Before it becomes entirely dependent on food manufactured through its true leaves, the root system develops, and the reserve food supplies in the cotyledons, or seed leaves, is exhausted. At this point, it can exist in a sterile growing medium like sphagnum moss or peatmoss, but once the true leaves appear, the plant will require a fertile soil or the addition of nutrients through fertilizers.

SEED DISPERSAL.

Nature has provided some marvelous methods to insure the scattering of seeds over the earth, thus increasing the chances of some seeds landing in a spot favorable for germination and survival until they, in turn, can send forth seeds of their own.

The maple (1) and the pine (2) have wings; the milkweed (3) and the dandelion (4) have parachutes of silky hair; the cocklebur (5) has myriad hooks enabling it to stick with great tenacity to you or your dog; the poppy (6) uses a stylized saltshaker; the coconut (7), with its many layers of loose fibers, can float for miles on the ocean waves, until it reaches a favorable shore; and unlike man, most animals and birds eat the seeds within the delectable fruits provided by the bounty of the fields, and these seeds are usually resistant to digestive fluids, and thus become transported to promising new spots for germination.

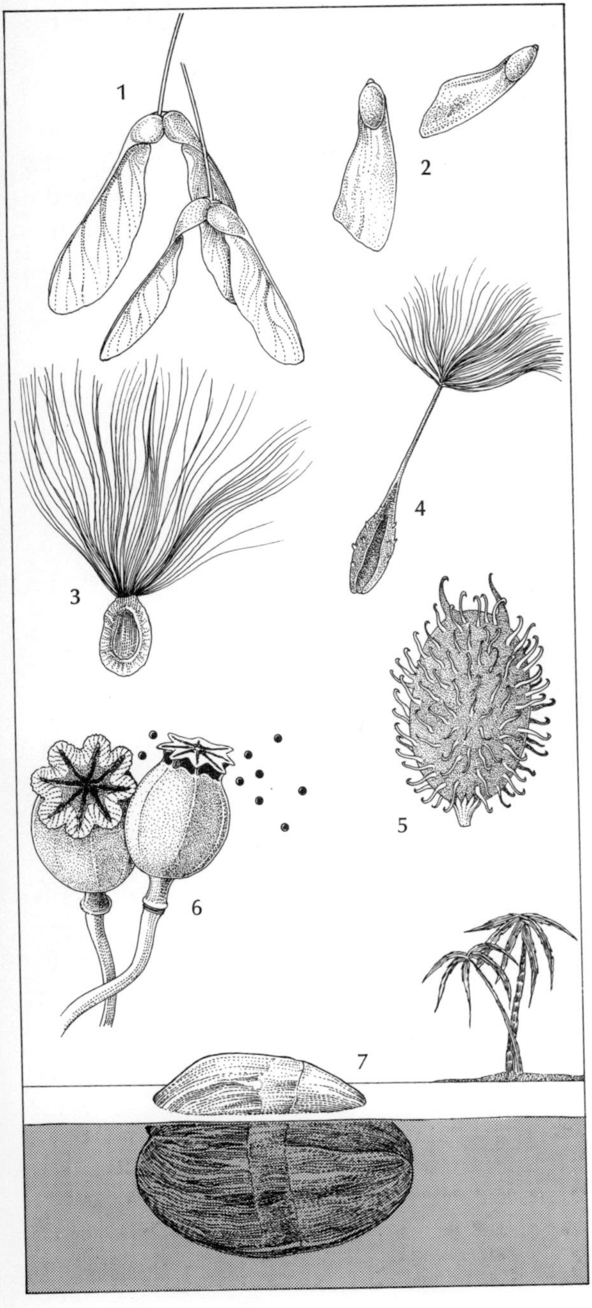
1
2
3
4
5
6
7

CHAPTER 2

Soils and Equipment

This chapter contains plans for a simply made seed-bed, a definition of the various materials used for growing mixes, and a small survey of products on the market today. All the products mentioned are available from the suppliers listed in the appendix (including electric seed-beds, if you don't care to make your own).

GROWING MIXES

"The medium is the message," and next to the seed, the most important item in propagation. The following materials are either used alone or in a combination with each other, and the mixes used for cuttings are made from many of the same ingredients.

- *Aquarium gravel and bird gravel*, found in pet supply stores and supermarkets, is useful in providing proper drainage to a mix, and is easy to find if sand, perlite, or vermiculite are unavailable. While it could be used alone for larger seeds, it would crush smaller varieties and contains no nutrients.

- *Humus* is the residue of the compost heap. It is well-decomposed vegetation that is black, sweet-smelling, loose in texture, but must be sterilized to prevent the worst threat to emergent seedlings: "Damping-off" (see chapter 3).
- *Loam* is the catchall word for good garden soil. It can be used but must be pulverized and sterilized.
- *Perlite* is made from volcanic rock, commercially cooked in giant pressure cookers. It is sterile. While performing the same function as sand, it's just too light to be used alone, and has no nutrients.
- *Peat*, sold by bale or bag, is the organic residue of many different kinds of plants that have decomposed in water. Slightly on the acid side, it's very difficult to get wet, and can easily pack down too tightly. It has some nutrients.
- *Peat moss* is partially decomposed sphagnum moss and, like peat and sphagnum, a chore to moisten if completely dry (use warm water) and eventually will form a surface crust that's quite impervious to water.
- *Sand* is one of the main constituents for a good mix, especially for cactus seedlings (see chapter 6). Use builder's or sharp sand, being sure to wash it well. "Sharp" simply means that the sand grains are rough to the touch, as opposed to soft sand, which is too fine to be useful. Beach sand is soft and contains salt, which is a disaster.
- *Sphagnum moss* identifies several different

mosses found in bogs and swamps. They grow very slowly, the lower parts gradually packing together to form a compact mass that is sold as peat. Sphagnum leaves are made of long, hollow cells with a fantastic ability to absorb water, and, coupled with their natural sterility (caused by a slight antiseptic quality), they make an exceptional medium for seed germination. Like all peats, they are initially difficult to get wet and if not kept continually moist, will form a tough, surface crust that seems to repel water. Sphagnum is sold both milled and unmilled. Either will work for germination purposes, but the milled sphagnum is easier to handle.

Sphagnum moss

- *Store-bought mixes*, available under many trade names, are generally very good, but much more expensive than making your own.
- *Vermiculite* is the lightweight material made from expanded mica, and often used for seedlings. It is sterile, does not pack down, but has a tendency to hold a little more water than the seeds demand. It has no nutritive value.

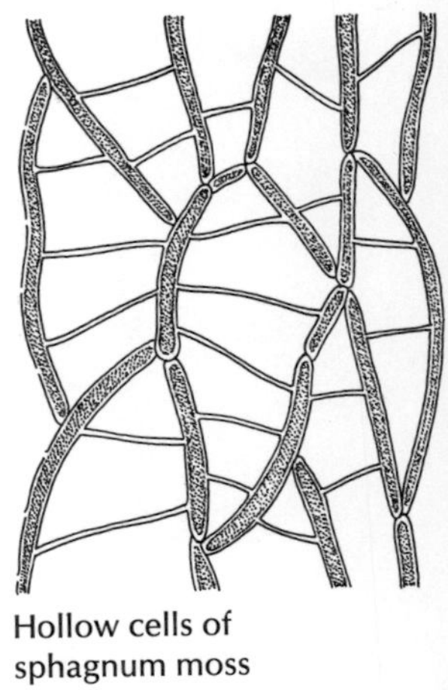

Hollow cells of sphagnum moss

Any of the materials listed above could be used to germinate seeds; after all, seeds fall in the most unlikely places and manage to survive. Nature can afford to waste thousands of seeds, but we can't. I've tried various combinations and found the best to be:

1/3 Milled sphagnum moss
1/3 Vermiculite
1/3 Sand

Since this is a completely sterile mix, we eliminate the problems of "damping-off"; it's very clean to use, so there are no problems

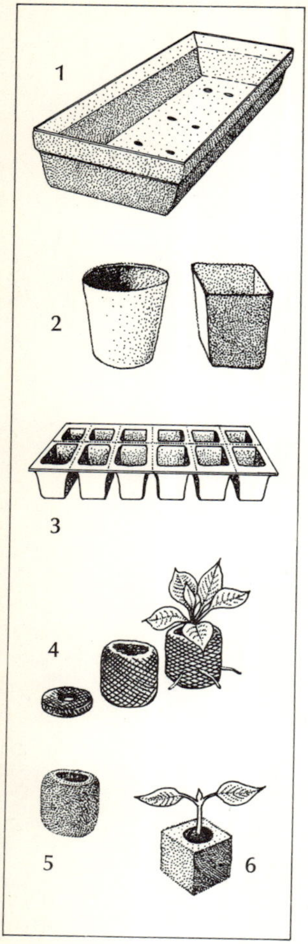

with dirt in the kitchen sink; and it's readily available in any garden supply store. Because there are no nutrients of any value to seedlings in this mix, fertilizers must be added when the true leaves appear, but more about that in chapter 4.

EQUIPMENT

One of the problems in our world of today revolves around affluence: too many cars, too many can openers, and too many types of equipment to grow seeds. Some, I admit, are a bit more aesthetic than old margarine cups, coffee cans, foil pans from frozen pies, and empty milk cartons, all of which make excellent seed flats, but it also costs a good deal more than using food packaging left over from last night's dinner.

So, for those readers with some spare cash and a thirst for the new, the following containers are available on the commercial market, either as individual items or components of kits:

(1) Flats are made of compressed peat, wood fibers, or plastic. They generally come in two sizes, 7½ x 5½ x 2¼ inches or 10 x 12 x 2½. Only use these for large sowings. (2) Peat pots both square and round are good for rooting cuttings and starting those seeds that resent transplanting later on. Remember when using these to slit the walls when repotting because some roots won't force their way through the peat walls, and never leave any of the top edge uncovered since it acts like a wick taking water from the soil and evaporating it into the open air. (3)

Peat pot strips are a series of smaller pots, molded together for ease of handling. They're great for the large-scale grower, but not necessary for the beginner. (4) Jiffy-7's are small pieces of compressed peat encased in a thin plastic net. When water is added, the discs expand to 2 inches high and 1¾ inches in diameter. Roots penetrate the open sides with ease. They are handy to have around to start one or two cuttings or planting just one or two seeds. (Although the advertising tells you of the fascination in watching them expand, you'll quickly get bored of that!) (5) Jiffy-9's work on the same principle, but without the netting, so they require more care in handling, or they'll crumble in your hand. (6) Fertlcubes are 1-inch square blocks of compressed sphagnum moss, perlite, and vermiculite with the addition of nutrients.

In addition, there are various combinations of the previous items, some with self-watering trays and wicks, marketed as complete kits for seed propagation.

Many stores now sell small indoor greenhouses made of plastic to insure the proper warmth and humidity. Although stylish, no doubt, and much better looking in the small apartment than old metal trays with odd-sized pots under saran wrap, they really don't work any better.

Just about any container that will hold the planting mix, that is fairly waterproof, has drainage holes on the bottom, and is at least over 2½ inches in height, will work for seeds. I don't wish to knock a troubled economy, but it's one of the cheapest hobbies around.

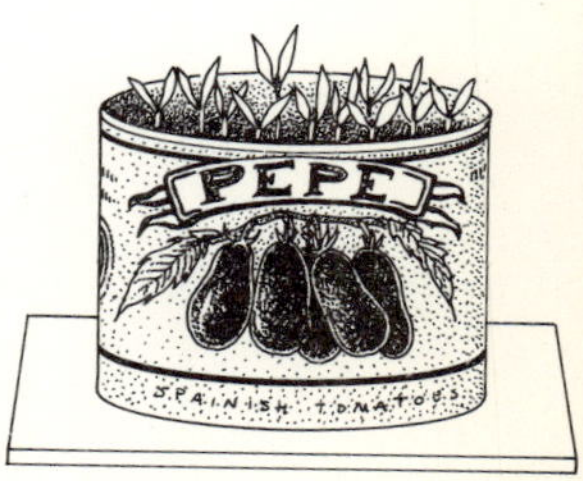

HEATING CABLES

The single most important reason that many people have trouble getting seeds to germinate is simply that the seed bed is too cool. While there are many crop plants that refuse to germinate unless subjected to temperatures below 15°C (65°F) most houseplants will not start growth unless they are above that temperature. Lack of heat works in combination with too much moisture, rotting the seeds before they can sprout. The majority of seeds need 18°-27°C (65°-80F).

Far and away the best solution is a waterproof, insulated heating cable. They're sold with or without thermostats. Pay a bit more for the thermostat (it's usually preset to 21°C [70°F]) just to be on the safe side. Cables vary in length from 12 feet to 120 feet in length. A 12-foot cable uses the power of a 40-watt light bulb, and at the time of this writing, costs six dollars. This will heat an area of four square feet. Remember, too, that most cuttings will root with greater efficiency when given the gentle warmth that a cable provides.

PROPAGATING FRAME

The following drawing shows a simply constructed seed-bed made from locally obtained materials:

(1) 3/4″ x 4″ planking (2) 1/2″ or 3/4″ waterproof plywood (3) Aluminum nails (4) Heating cable (5) Thermostat (6) Heavy plastic sheeting (7) Staples (8) Sand or gravel filler

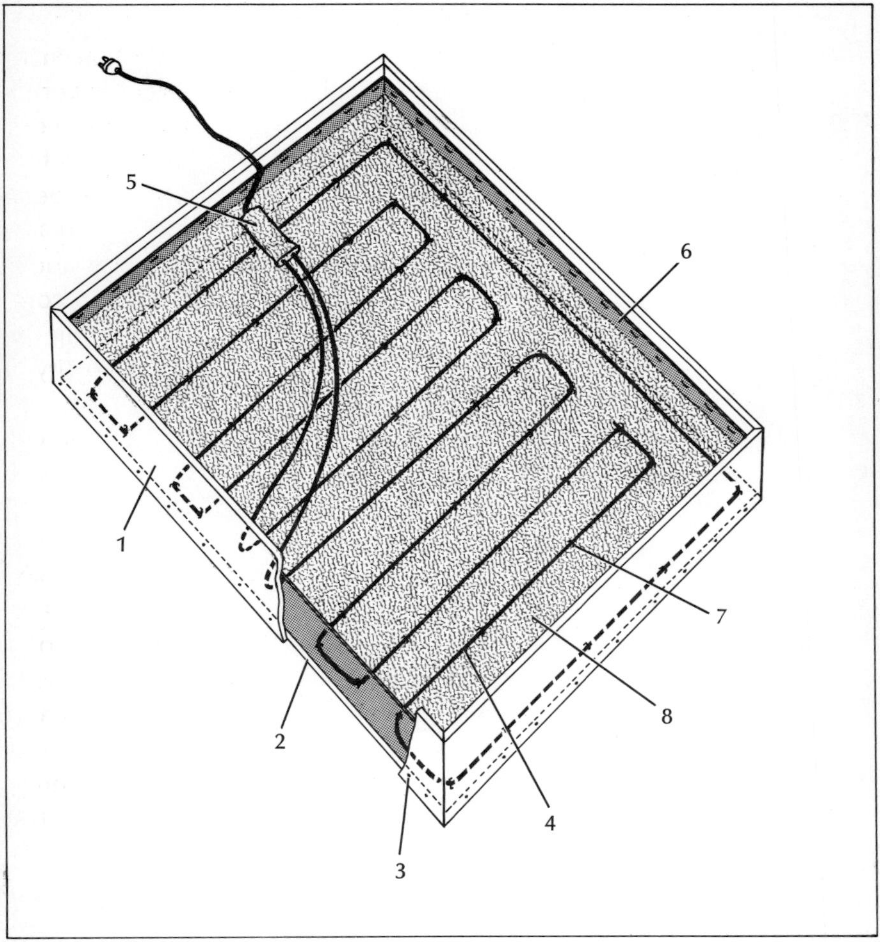

The cable can be held in place with staples —just be careful not to pierce the plastic insulation. Place the outside rows of the cable 2 inches from the edge of the box. Leave about 3 inches between the adjacent rows. Don't let the cable touch the thermostat, as it will shut off before the soil is properly heated, and don't overlap or cross pieces of cable, since it will prevent even distribution of heat.

CHAPTER 3

Success with Seeds

It's always more enjoyable to grow your plants from seed; there's a certain excitement in seeing the first touch of green, pushing up through the soil, and knowing that you provided the necessary environment that enabled the seed to germinate. When mature, you can take full credit for that lovely plant, not the professional grower.

THE BEST SEASON

Try to buy and start your seeds in early spring. Take full advantage of the long growing season ahead. Although some plants may be initiated at any time of the year (see appendix), the majority profit from matching their growth with the lengthening days of spring.

Try not to be smitten with the lovely pictures in the catalogues. Limit your buying to what you have room and equipment for. Remember that your time is valuable and the sight of endless trays, all with emerging plants, can be an awesome responsibility.

STORING SEEDS

When seeds arrive, you'll notice that most are packed in foil packets, with instructions for sowing on the label. These packets are airtight and sealed under a vacuum, keeping out light, air, and moisture: three of the four musts for germination. To insure their continued vitality, store in a closed jar and keep in a cool spot away from radiators. Never open the packets until ready to use.

While most seeds will grow after their second year of storage (and beyond), some lose vitality, especially when left to lie about in the open air. Instead of getting up to 100 percent germination, you'll wind up with 20 percent or less. It's also a good idea to mark each packet with the date of receipt, since it's very easy to forget.

Check to see if any seeds, whether purchased or collected, need stratification. If indicated, now's the time to start that procedure, by opening the package and placing the seeds in a slightly moist mixture of sand and/or sphagnum moss, placing in a "baggie" with the top tightly sealed and labeled as to date, and storing in the refrigerator for at least two months. When the right time has passed, the seeds and medium can be sown as a unit.

Some seeds take a *very* long time to get started: bananas, lantanas, and palms may lie dormant for a year or even longer, so you've got to be prepared to cope with such slow-starters.

It is wise to make a growing chart with the day of planting, the popular and scientific name, and the projected time of germination. The Park Seed Company publishes an excellent two-page germination guide in their popular catalogue.

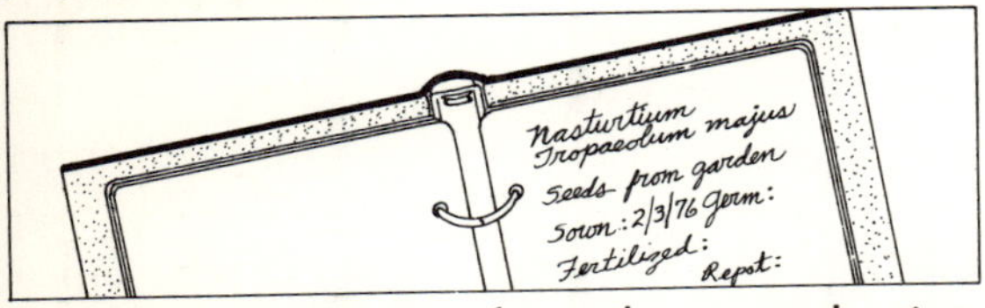

Choose a container from the vast selection open to you, being sure that drainage holes are in the bottom (or if using a solid container, spread 1 inch of gravel on the bottom for excess water). Check to see if any particular plant resents transplanting at a later date; morning glories and nasturtiums are famous for this. For these plants use a small container like a peat-pot, Jiffy-7, or a 1/2 pint cream or milk container. When seeds are very tiny, use a larger container so you will end up with a more even distribution.

To avoid identification and transplant problems later on, never put more than one type of seed to a container. Keep some seeds from the package for later sowings, in case an unplanned-for emergency arises.

PREPARING THE MIX

Now using materials from chapter 2, you're ready to prepare the growing medium. If you're going to utilize garden soil to form any part of the mix, sterilize it by baking in a 250°F oven for at least two hours. This is to prevent "damping-off," a disease of seedlings caused by fungus often associated with unsterilized earth and encouraged by the moisture in seed germination mediums. It's the main reason for using sphagnum moss and other man-made materials. There are chemicals, like Benomyl, that will kill any fungus spores, but if your mix is clean to begin with, there should be no problems. "Damping-off" is not to be passed over lightly; you go through a great deal of time and effort raising these seeds, just to see them fall over and wilt, and at that stage you can do nothing to cure the problem.

Fill your containers close to the top, leaving 1/4 of an inch or less so that air will easily circulate, tamp it down and proceed to wet it thoroughly. Pour water over the top using only mild force, as this helps to settle the mix. Watering from the bottom should be avoided until the top is wet, since the very light and dry materials will simply "pop" out the top. Now, drain well for at least an hour.

Another way is to put your ingredients in a bowl, add water, stir well, and squeeze all the excess water out with your hands.

We know that seeds need water for germination, but they also need oxygen from the air for life processes to begin. If we are too heavy-handed, the mix will be so wet that the seeds will be completely surrounded by water, *and no oxygen* will be readily available to the embryo plant and it will die. This is the reason why control of watering is so important.

If you are worried about how deep to plant the seeds, if you have collected the seeds yourself, or if the seed package neglected to give the information, use the following rule to determine the depth at which seeds should be planted: seeds, 1/16 of an inch, or larger, should be covered by the thickness of one seed; tiny seeds, like begonias, need not be covered at all, but just settled in with a light spray of water from a hand mister. When sowing smaller seeds, use a folded piece of paper, tapping it gently as you move it across the surface.

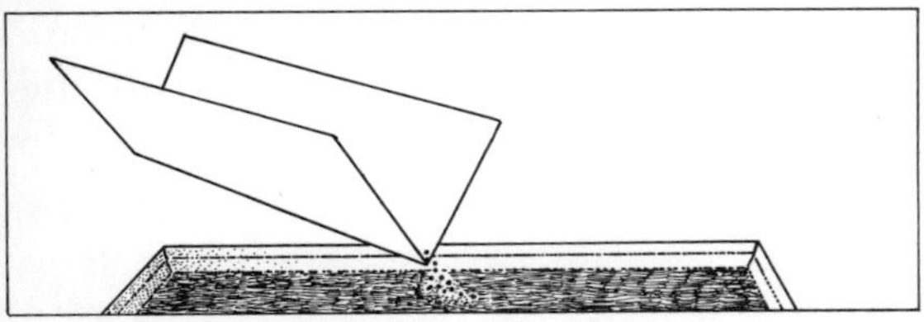

Now that the seeds are planted, cover the container with plastic wrap or glass. Be careful to keep plastic wrap away from the surface; a plant label or small stick can act as a tent pole. Drops of moisture will condense (especially if you are using a heating cable) and could form small swamps that will drown the seeds. This miniature greenhouse will prevent the medium from drying out—an event to be avoided at all costs; once germination begins, smaller seeds, especially, must never be allowed to dry out. Put the containers in your propagation box, or in a warm spot away from the direct rays of the sun. Soon they will want all the sun available, but not until they sprout. You can even use an electric light of about 60 watts inside a box to generate heat.

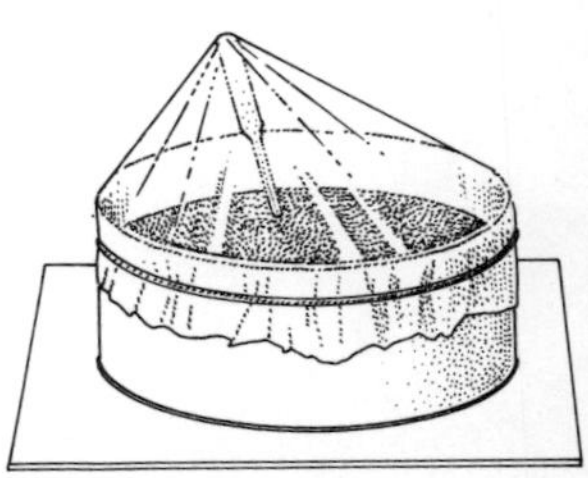

A lot of sources will tell you that complete darkness is necessary for germination, but I never found that to be the case. In fact, many seeds need the stimulation of light before they can start the complex chemical activities needed to begin growth. A few varieties must have total darkness, like calendula and the cyclamens, but the seed packet will point that out.

When the first green shoots appear, move the containers into the direct sunlight (unless you're doing this during late spring or summer, when caution must be taken with the very hot rays of the midday sun; some protection like screening should be used). Turn the plants daily to keep them from bending to the light. Remove the plastic or glass covering and check on the water situation at least twice a day. When the mixes start to dry, either water from the bottom through the drainage holes or from the top with a mister, or the fine rose on a watering can.

After the true leaves appear, it's time to use a liquid fertilizer, like Hyponex or Miraclegrow, to make up for the lack of nutrients in the mix. Follow instructions, then dilute the solution by half. Even if you are using soil, the applications of such a solution will not endanger the plants.

You can now transplant to small peat-pots or if leaving the plants in their original flats or containers for a while longer, thin them out to at least 1 inch between each plant. These small plants will have to remain in the original container until large enough to handle.

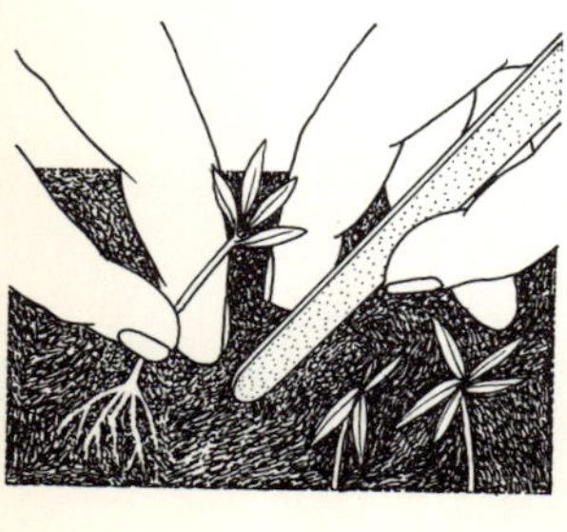

The pointed end of a knife or a plastic plant label makes an excellent tool for transplanting. Just be gentle! Pick up the plant and lightly cover the roots with soil that is damp, but not mucky.

The plants should not be moved into 3-inch pots until they are about 3 inches high.

FIVE RULES FOR PROPER GERMINATION

1) Don't overwater, since you deprive the seed of needed oxygen.
2) Never let the mix dry out.
3) Use a mister or the fine rose on a watering can.
4) Never sow seeds too deeply. If in doubt, just cover them lightly.
5) Keep seeds from sunlight until germination is complete.

COLLECTING SEEDS

Most people dislike plant-nappers, whether in the home, the park, or the greenhouse, but few are the avid gardeners who won't give away a few seeds (or cuttings for that matter). On your travels, keep an eye out for seeds, remembering that the plants of the southern part of the United States become houseplants in the north. Ask friends who travel to remember you, and instead of sending postcards, send an envelope of seeds. One of the happiest Christmas presents in my memory came from my sister-in-law, in Seattle, Washington, who sent a box of many seeds, each marked with the date of collection and the place. It kept me entertained long after the memory of a tie or gadget would have faded.

CHAPTER 4

Success with Cuttings

If buyers had to wait for suppliers to grow all the plants we desire from seed, there just wouldn't be enough to go around. Some plants take years to reach maturity, and you would be providing "bed and board" for all that time. In addition, many that you will buy today are hybrids, results of cross-fertilization, and thus their seeds will produce plants with flowers quite different from those of the parent. Think of the variations found in hair and eye color among people (who are hybrids) and you have a glimpse of the problem. Then we have plants like the edible banana, and many houseplants, which never produce viable seed, but must always be grown from cuttings, year after year. Where do they leave the propagator?

Luckily, nature has accomplished an alternate method called asexual or vegetative reproduction, a process that gives most plants the ability to grow mature offspring from special cells found in leaves, stems, and even roots. One plant can manufacture hundreds of offspring over its lifespan, each identical to the parent, and all bearing the look that only instant maturity can give. It's a natural process that nature has used for millions of years, and mankind since the 1700s.

A dendrobium orchid, for example, comes from an area of the Himalayas where spring and summer are warm and lush while fall and winter stay cool and dry. In normal weather, this orchid loses its leaves in fall and brings forth magnificent flowers in spring that mature and go to seed. But should the fall and

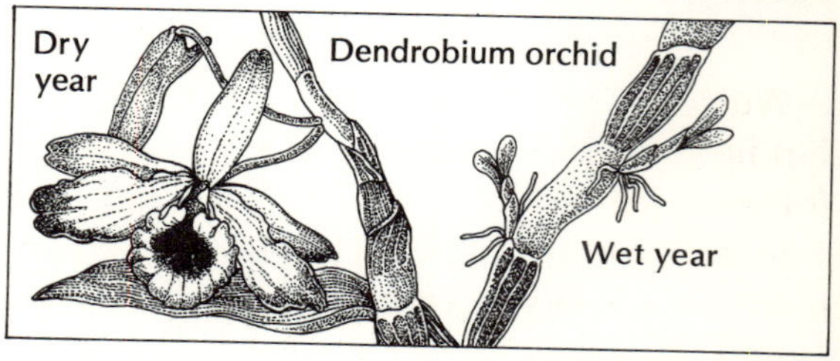

winter be inordinately wet and warmer than usual, the plant makes miniature copies of itself complete with roots, forgoing flowers for the coming season. The dendrobium has evolved so that it has a second chance: in those years when triggered by bad weather, it produces more complete plants in one year than it ever could by means of the regular three-year cycle.

By cutting, layering, and manipulating a plant's environment, we can do far better than the dendrobium orchid.

The following chart shows the different parts of a plant that we utilize in asexual reproduction:

LEAF CUTTINGS
- With petiole (or leaf stalk)
- Without petiole
- Severing veins
- Sectioning leaves
- Leaf bud

STEM CUTTINGS
- Hardwood
- Semihardwood
- Softwood
- Tuber
- Rhizome

ROOT CUTTINGS

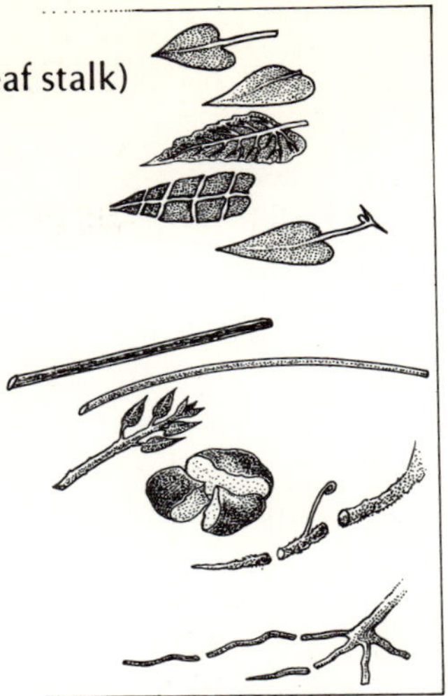

We'll go over each of these, step by step, but first a bit of information about containers for propagation.

In addition to the time-honored clay pot, there are varying types of peat and moss containers, as described in chapter 2, that are very handy for cuttings, and many more commercial designs have appeared on the market in recent years. They are all based on the idea of a clear plastic breadbox with drainage holes in the bottom and ventilation holes in the top. The suppliers listed in the appendix carry many of these items, and while I believe in making do with what you have, propagating boxes are neat, compact, and efficient.

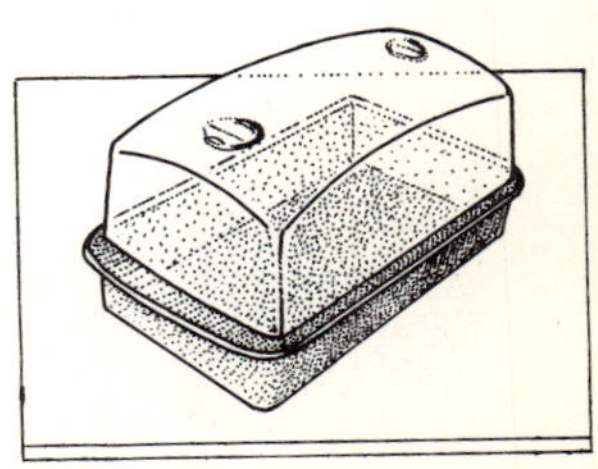

Since most cuttings respond to gentle bottom heat, the propagating box, described in chapter 2, is extremely useful for cuttings. While requiring some effort in its manufacture, you'll never find it a waste of time or money.

PLANT HORMONES

Plant cells resemble animal cells in many ways, both using hormones to control and stimulate growth and development. Often, a cutting that refuses to respond to the rooting process simply needs the additional stimulation of a plant-growth hormone. There are three types on the market: Hormodin and Rootone, both for softwood cuttings and Rootone #10 for hardwood cuttings. The wound on the bottom of the cutting is dipped into the powder, just prior to planting.

SOFTWOOD STEM CUTTINGS

Of the five types of stem cuttings mentioned, the one most often used is the *softwood*. The term refers to the soft, green, stem tissues found in houseplant annuals, or the first season's growth of perennials before the stem matures into woody tissue.

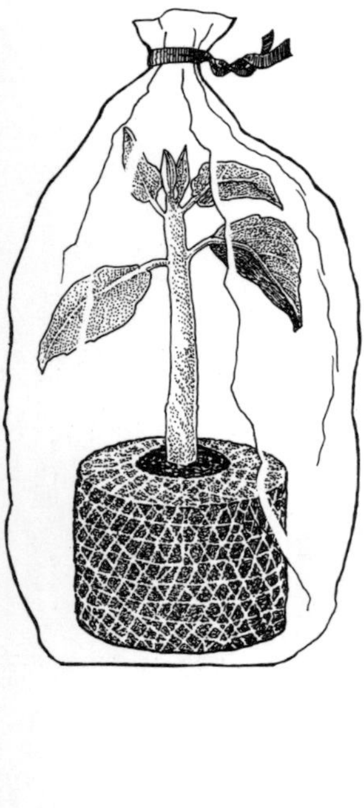

Late spring or summer is the best time to start cutting since at this season the plant has a good supply of food stored in its cells to keep it going until new roots form. Most plants will root at other times, but the rooting process will be slower. It has also been scientifically demonstrated that the younger the stem, the faster new roots will appear.

If only a few new plants are wanted, I've found the best method is a combination of Jiffy-7's and small plastic food bags. First, choose a plant that needs pruning. Cut the healthy stems, which are about 3 to 5 inches long, with a sharp knife or razor blade slightly below the point where a leaf petiole (or stalk) joins the stem. Remove any damaged leaves and flowers and neatly slice off any bottom leaves close to the stem. Take an expanded Jiffy pot (any other small container will do) and make a hole with a pencil or similar object, 3/4 of the pot's depth. Insert the cutting, making sure that the base of the stem touches the bottom of the hole. Firm the medium back into place around the stem. Now put the whole affair in a "baggie" with the opening at the top.

If the air is very dry, seal the bag with a twist. The baggie holds the moisture that the leaves throw off, but cannot replace until new roots form. The medium should be moist at all times, but never soggy! In about two weeks, give the cutting a slight tug to check if rooting has commenced. If not, pull out the cutting and see if the end has started to rot. If all looks well, try again, perhaps dusting with one of the hormone powders. Make doubly sure that the base of the cutting is touching the rooting medium; it needs the stimulus of this contact to start new roots. When new roots appear, poking through the mesh, plant the Jiffy-7 in a 3-inch clay pot.

When you wish to root many new plants, use a 6-inch clay pot filled with milled sphagnum moss, or some of the mix left over from seed propagation. It should be evenly moist, not sopping wet, and tightly packed. Cuttings are inserted and the pot placed in a plastic bag. If the humidity is low, or the plant leaves thin and fragile, close the bag. Plastic bags that are sealed should be opened every few days and checked for mold. The stagnant air and damp conditions give mold and fungus a helping hand.

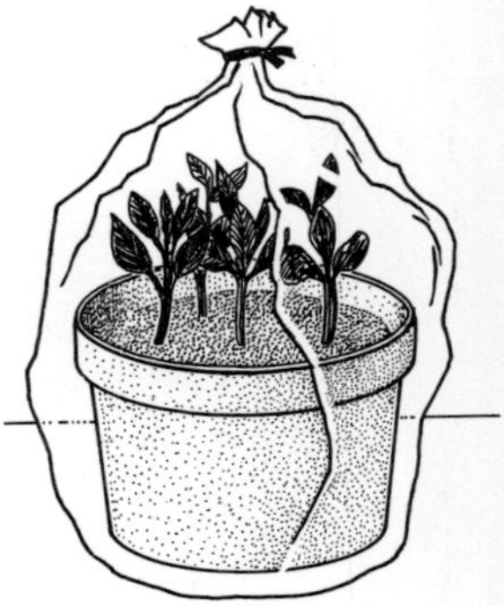

SEMIHARDWOOD CUTTINGS

Many plants like fuchsias, camellias, and gardenias, are multiplied by taking cuttings from shoots of the current growing season in the fall of the year. The stems selected should not have hardened into wood, but bend easily without snapping or pulling apart. Cuttings should be 4 to 6 inches long, with the soft tip removed, the bottom being cut just below a node. Place the cuttings in a pot with moist growing medium, or a frame, with about 1/3 of the stem buried, making sure that the base is in contact with the medium. To be doubly sure of rooting, dust with rooting hormone. Cover the pot with a plastic bag and subject to gentle bottom heat of 18°-21°C (65°-70°F). The cuttings should begin to root within a month, when they can be transplanted to small pots.

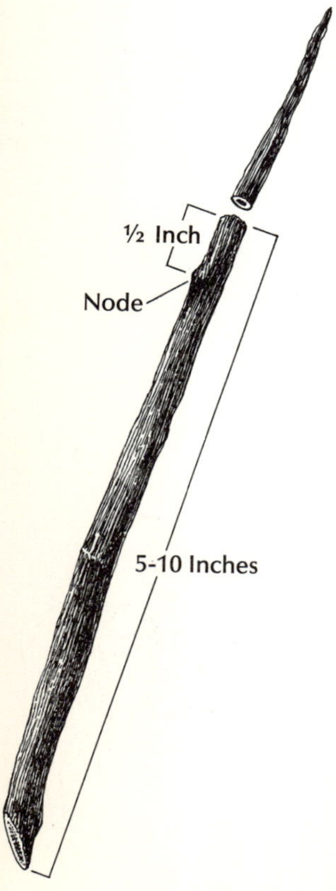

HARDWOOD CUTTINGS

Hardwood stem cuttings are made after the soft stem tissue has actually hardened into wood. Growth from the previous season is used. This method is usually associated with outdoor flowering trees and shrubs, but I include the directions just in case you live in the southern parts of America, where softwood cuttings have a greater tendency to rot before they root. The method also works well for citrus fruits and poinsettias.

Sections should be at least 5 inches and up to 10 inches long. The bottom cut is made at a slant (so you know which end is up when the twig is bare) just below a node and the top cut about 1/2 inch above a node. The tip of the stem is generally discarded since that part is usually low in stored food. The bottom is dusted with a hormone powder and planted in the growing medium, at least 2 inches buried for three to five weeks with a gentle bottom temperature of 18°-21°C (65°-70°F) with the tops exposed to the open air. As roots develop, the stems should be transplanted into pots.

TUBERS AND RHIZOMES

Although there is no resemblence, tubers and rhizomes are really modified stems. The most familiar tuber is the potato, but gloriosa lilies, Gloxinias (*Sinningia* spp.), and tuberous begonias are also examples of this group of plants.

To multiply tuberous plants, the tuber may be divided into as many pieces as there are eyes (or axillary buds). Use a sharp knife, and set the cut pieces in a warm, dry place (20°C, 68°F) for a day or two, allowing the cuts to heal. Plant the pieces in a moist mix, with just the eye uncovered. When new leaves and roots appear, move to individual pots.

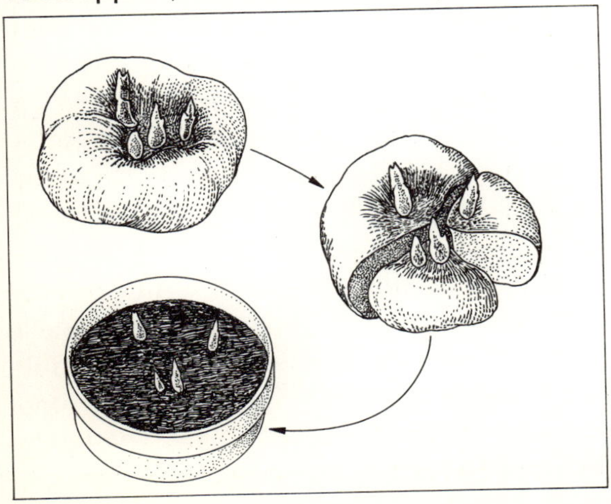

Rhizomes are represented by the well-known iris, the sweet flag (*Acorus* spp.), and many ferns.

If a rhizome is cleanly cut into sections, each piece with a node, the cutting will form new plants. Lightly cover the pieces with moist medium and transplant when new roots and leaves develop.

LEAF CUTTINGS

There's a limit to the number of new plants to be grown from stem cuttings, but the amount of progeny from leaf cuttings of one type or another is fantastic; armies of plants can be grown in this fashion.

Leaf Cuttings with Severed Veins

This method is used with *Begonia rex* varieties. A large, healthy, and mature leaf is chosen and the periole shortened to the length shown. The main veins are then cut at vein junctions, using a sharp knife or razor blade. The leaf is then laid on a moist bed of mix, top

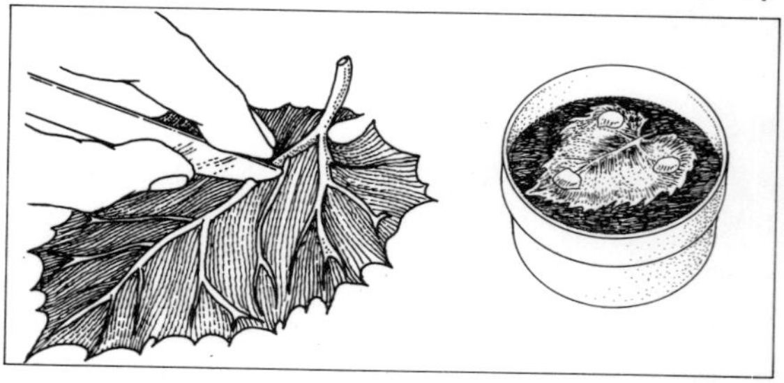

side up. The petiole is stuck into the rooting medium and the leaf held in place with small stones or bobby pins. Cover the pot with plastic and keep it out of intense light. Within a few weeks, miniature replicas of the parent plant will sprout on the leaf surface.

Sectioning Leaves

Another way to quickly multiply *Begonia rex* and the other large-leaved begonia varieties is with leaf sectioning. Cut the selected leaf into small squares. Each piece must have a main vein running through it. Follow the rooting procedure for the previous method.

Mother-in-law-tongue (*Sansevieria* spp.) is another common houseplant that may be reproduced by sectioning the leaf. Cut a healthy leaf into pieces, each 3 or 4 inches long. This process will not work if the leaf is upside down, so make the bottom cut on a slant just to keep everything straight. Plant 3 inches of the cutting into a rooting medium,

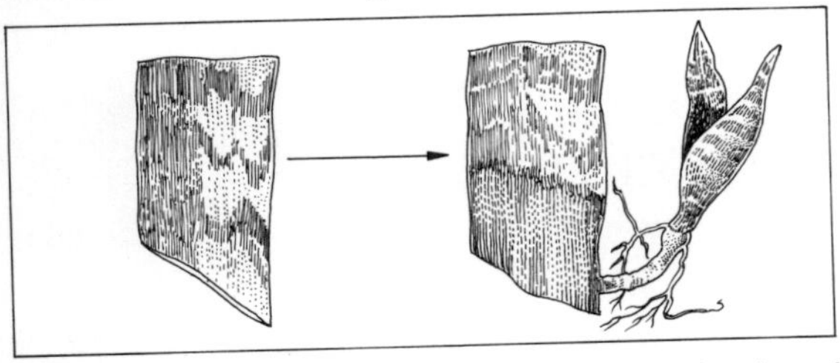

and a new plant will develop to one side of the mother leaf. An interesting thing to note is that a variegated leaf will produce an all-green plant, since the genetic variation that produces the variegated form occurs only on the surface of the leaf, and the new plant shoot and roots arise from the inner, or non-mutated core.

Single Leaf Cuttings

Streptocarpus, gloxinias, African violets, peperomias, hoyas, and many other plants can be reproduced by using variations of a single leaf, leaf with petiole, or the leaf, the petiole, and a section of the stem including a leaf bud.

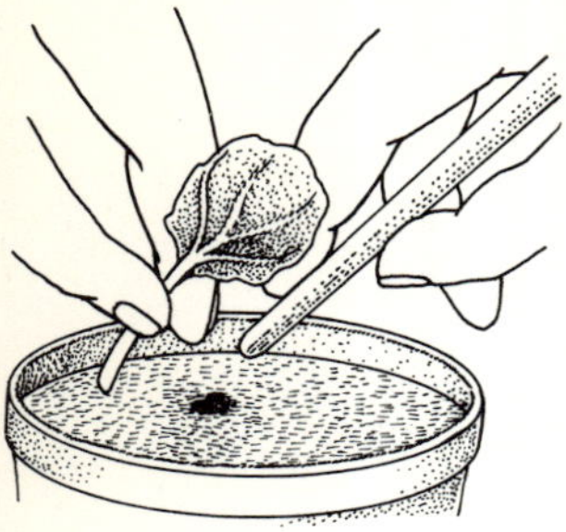

Leaves that are thick and succulent can be pushed into a moist medium without a petiole. Kalanchoes, and other succulents, will root and form new plants.

Leaves are pushed into the medium with the petiole attached. This method works on African violets, most begonias, hoyas, and others.

If either of these procedures doesn't work, try including the leaf bud and a bit of stem. The jade plant (*Crassula argentea*) will form roots from a leaf alone, but it will never grow a stem unless the leaf bud is included.

When using leaves that are thick, waxy, and of a succulent character, the cuttings need not be covered, but be sure the medium stays moist.

ROOT CUTTINGS

Plants with thick, fleshy roots, like clivias, bananas, plumbagos, and clerodendrum can be reproduced by cutting roots into 3- to 4-inch sections and laying the sections, lightly covered, on a moist bed of mix. Bottom heat should be applied with the heating cable. Make these cuttings in the fall or early winter to take advantage of the stored food in the roots.

OTHER METHODS OF PROPAGATION

Here are five more methods to multiply your houseplants:

Cuttings in Water

An old-time method is rooting cuttings in a glass of water. It works for ivies, impatiens, coleus, begonias, wandering Jews, geraniums, and many more. (The chart in the appendix marks plants that grow roots this way with a W.)

One thing to remember is to change the water every few days, to give new oxygen to the plant. If water remains unchanged, the rooting will usually occur near the water's surface, where oxygen is more available, and there won't be enough new growth to support the plant.

After roots have shown healthy growth, add small amounts of sand or soil to the water every few days. You won't have a tangled mass of roots to sort out when ready to transplant, and the soil or sand will stimulate the roots to grow new root-hairs, which are the true providers of food and water to the plant.

Plants by Division

Another easy way to produce new plants is by division of the root stock. Early spring is the best time, before new growth for the coming season begins. This process works with plants that form multiple crowns, like many of the ferns, and clivias, aspidistras, asparagus fern, lily-of-the-Nile, and a host of other houseplants; it's obviously a bad idea with a single-stemmed plant. You can divide into as many sections as there are individual stems or crowns.

Try to separate the roots as carefully as you can, keeping as many as possible attached to their original stem.

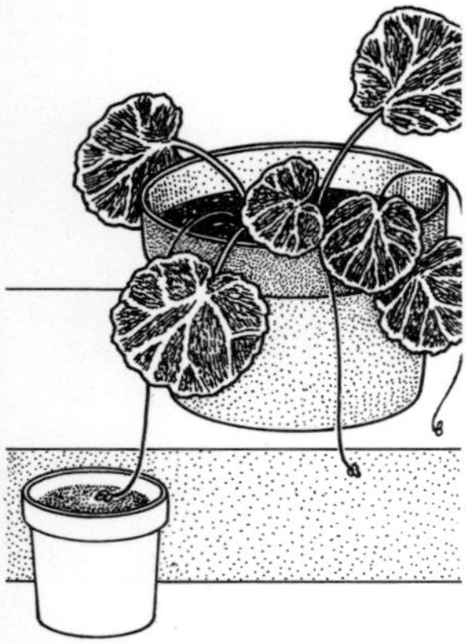

Runners

Any houseplants that produce runners with tiny plants at their tips, as strawberry geraniums and spider plants (*Chlorophytum* spp.) will make new plants if the tips are allowed to root. The easiest way is to anchor the small plantlet to a 3-inch pot of soil using a bobby pin or a paper clip. Make sure that adequate roots have formed by giving a slight tug before you cut the runner from the mother plant.

Layering

When a prized dracaena or rubber plant (*Ficus* spp.) has reached the ceiling, it's time to watch it bend or air-layer it. This is an especially valuable method to produce a mature plant in a short time and relieve that leggy look.

Start the procedure in spring by making a diagonal cut, about 1½″ long, in the stem at the height you wish the new plant to be. You'll have to remove a few leaves above the cut. Keep the cut from healing by inserting a wooden matchstick into the wound. The cut area is then surrounded with thoroughly moistened peat or sphagnum moss and wrapped with plastic, tied top and bottom. Inspect the package every week or so for signs of new, white roots. When roots are well developed, remove the plastic, but leave the moss and cut the parent stem cleanly just below the new roots. Pot in a 5- or 6-inch pot and be very gentle as the new roots will be quite brittle.

By removing the top, the remaining stem should start to produce new leaves.

Cane Sectioning

Dracaena, dieffenbachia, and cordylines may be propagated by cutting their stems into 3-inch sections, making sure that each section includes a leaf scar, then placing them horizontally on a bed of moist sphagnum or peat moss. When rooting is well along, the dormant bud will sprout and a new plant is on its way.

BULBS AND CORMS

Although bulbs and corms are modified stems (like rhizomes and tubers), I've listed them separately, since most people think of them as entirely unique structures.

Most can be grown from seed, but take anywhere from two to four years before reaching the flowering stage, so asexual methods of propagation are especially valuable with these plants.

Division

Most bulbous plants increase their underground size year by year by producing offsets and eventually fill a pot, completely exhausting the soil. When plants reach this stage (or before, if you remember), remove it from the pot and divide the clump into smaller groups, being careful to avoid injuring the bulbs.

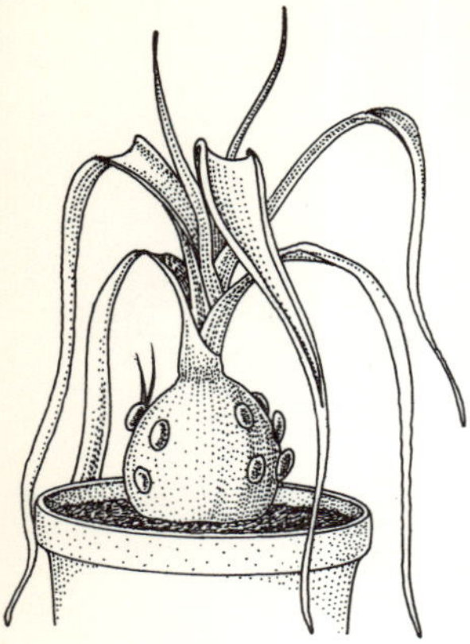

Bulblets and cormels

Some bulbs and most corms produce small buds on their underground surface (the false sea-onion [*Ornithogalum caudatum*] is the exception that proves the rule, and goes through the process above ground) called bulblets and cormels. These will be found when you are involved in dividing plants.

The bulblets are usually found in the lily family, and form in clusters about the stem, just below the ground. These may be carefully removed and planted in small pots, and covered with soil to the depth of the thickness of one bulblet. They will usually take three years to reach flowering stage.

Cormlets cluster around the base section of mature corms, where they are easily "popped-off" to be planted just like bulblets, where they will then flower in one to two years.

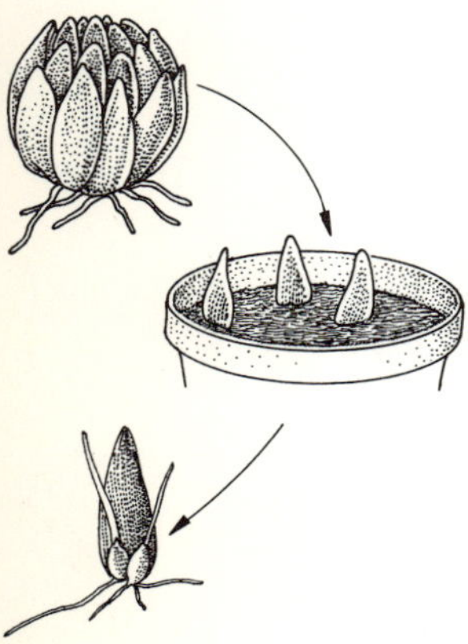

Propagation by Scales.

This process will work for most any lily species and consists of removing complete, individual scales from the bulb, rooting them in one of the propagating mixes at a temperature of 70°F (21°C). Stand them upright, making sure that the base is covered by the mix, and new bulblets should form within three months.

Basal Cutting.

This method is used primarily with hyacinths and is usually reserved for commercial growers. I've tried it with a hyacinth, and it certainly works, but I personally do not know if it will succeed with other bulbs, so experiments seem to be in order.

Mature bulbs are used, after the current foliage has died back. The basal plate is scooped out with a rounded spoon, then the bottom scored with three incisions, crossing in the center. The cuts are dusted with powdered charcoal and/or sulfur, then set aside in a dark place allowing a callus to form over the wounds. After this, the bulb is planted in the usual manner and bulblets should form around the cut areas.

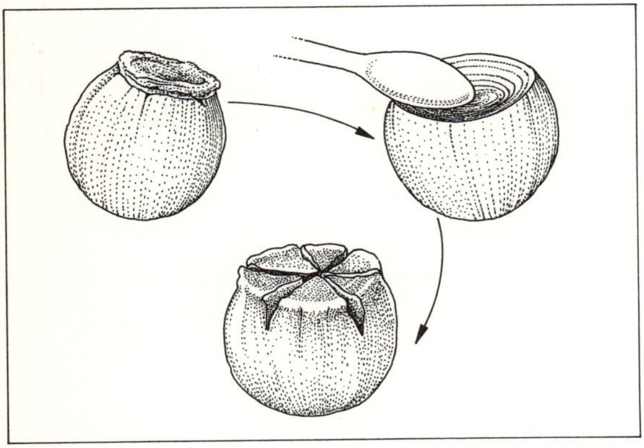

Leaf Cuttings.

The African blood-lilies (*Haemanthus* spp.) and the Cape cowslips (*Lachenalia* spp.) are two popular, bulbous houseplants that may be reproduced by leaf cuttings. Remove an entire leaf as close to the bulb as possible, placing the lower end in rooting medium to a depth of 2 inches, and providing bottom heat of about 70°F (21°C). Keep the humidity level high to prevent drying out. Roots and small bulblets will form within a month, which can then be potted up. It's entirely possible that this method will work with many bulbous plants, but little research has been done.

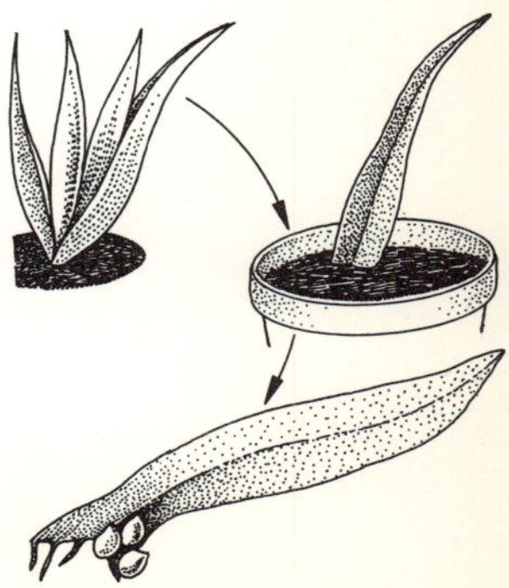

CHAPTER 5

Growing Ferns from Spores

In Shakespeare's *Henry IV*, Gadshill, a crony of Sir John Falstaff, explains to the court chamberlain: "... we have the receipt of fern-seed, we walk invisible." To which the chamberlain replies: "Nay, by my faith, I think you are more beholding to the night than to fern-seed for your walking invisible."

Thus, Shakespeare knew of the myth surrounding fern reproduction: they were popularly thought to produce an invisible seed, and anyone who possessed the magical "seeds" would be blessed with invisibility. He might also have been an amateur botanist, since he expresses a measure of mystery in the process.

It was not until the early 1800s that the mechanism of fern reproduction was finally resolved and the true "fern-seed" discovered.

LIFE CYCLE OF A FERN

Ferns do not bear flowers and seeds but begin their life cycle by producing spores* within cluster of spore cases called *sori*. Sori are found on the undersides of fern leaves and are often mistaken for insect infestations or disease.

The spore cases, at first green, turn brown as they age. When mature, and weather conditions are fairly dry, the walls of the sori bend back and discharge literally thousands of dustlike spores into the air. If it's raining, the sori remain closed, protecting the spores from destruction by heavy drops of water. The spores drift to the earth, sometimes traveling miles on the breeze, and with proper conditions of temperature and moisture, grow into flat, heart-shaped structures called *prothallia*. At maturity, prothallia are about the size of a green aspirin tablet and

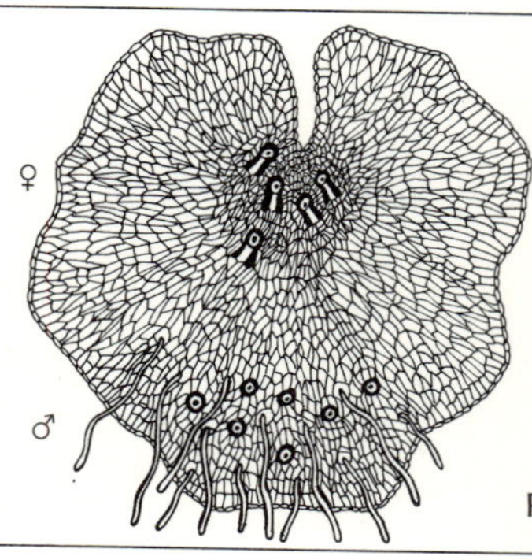

Prothallus

*Spores are reproductive bodies formed by ferns and algae, lichens, fungi, and mosses as well. They usually contain but one cell and are only one of the many steps that lead to a new plant. Seeds are produced only by conifers and flowering plants and contain an actual embryo plant of many different cells and complicated structure. Seeds grow directly into adults upon germination.

bear no similarity to a traditional fern frond. Separate and distinct male and female organs soon grow on the underside of the prothallia, producing eggs and sperm. With water from dew, rain, fog, or even melting snow, the sperm swim to the egg and fertilize. In a short time an embryo fern begins to develop.

Although the fern's life cycle sounds a bit complex and chancy, they are very easy to grow from spores. Unlike flowering plants, all ferns have the same general requirements for germination.

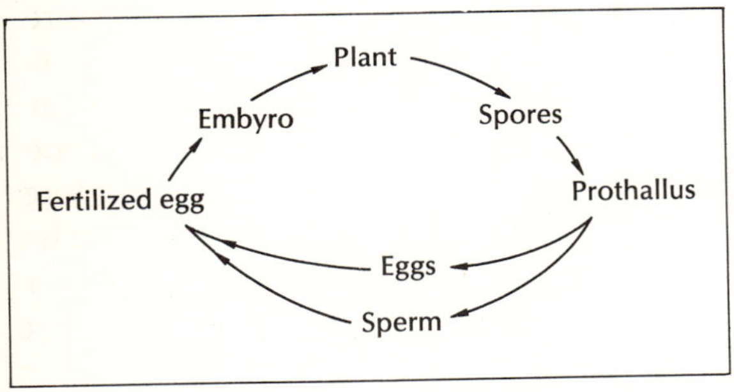

PROPAGATING THE SPORES

Absolutely sterile conditions are mandatory. The growing conditions for fern spores match those of mold, fungi, and a host of other undesirables. Use 3- or 4-inch clay pots (not plastic), scrubbed and clean. Crock the pots and fill to 1 inch of the rim with a sterilized soil mix of 1/3 potting soil, 1/3 peat moss, and 1/3 sand. The growing mix should have adequate drainage, so don't omit the sand. Tamp the mix, making it smooth and level. If you're unsure about the sterility of the mix, put the pot and soil into a 250°F oven and bake for two hours. Although it all sounds a bit like a

TV hospital drama with all the emphasis on being clean, it's very important. There's no sense in setting everything up to be destroyed by a wayward fungus.

Now soak the pots and mix until they are completely saturated; then drain off the excess water.

Cast the spores at any time of the year, one species to a pot, using the folded paper method from chapter 3. Label each pot with the date and species of fern. Baby ferns all look alike, so labels are necessary. Cover the pots with a sheet of glass, or unbending plastic, and put them in a warm spot at 65°-70°F (18°-21°C), with dim light. At all times, keep the pots away from the direct rays of the sun, as this will impede germination. Check the medium for signs of drying out. If it starts to dry, soak the pots. Don't water from the top because germinating spores should never be disturbed. If condensation on the glass becomes too heavy, remove the glass for a while. A thin beading of water is what we want, not huge drops of rain.

Many species will germinate within a few days, others will take a few weeks, so be patient. As the prothallia grow, a green cast will appear on the surface of the mix. In about three months, the prothallia will reach their full size. If you have achieved excellent spore germination, thin the prothallia out so they're about 1 inch apart, since only the male organs will usually develop if the prothallia are too close together. Falling condensation from the glass should provide enough water for the sperm to swim to the egg, but you might have to mist the surface, if it shows signs of drying.

The tiny fern plants will now begin to develop, growing from the center of each prothallium. As the fronds grow larger, you'll have to thin the plants again. When three or more fronds have appeared, transplant to individual pots. It will take about two years for a mature plant. You'll now have a more-than-adequate supply of plants for house and garden, since the same methods may be used for expanding your collection of native, outdoor ferns.

When growing ferns in pots or terrariums, it is not at all unlikely for germination to take place entirely without your help or knowledge; baby ferns will pop up in the most unlikely places. I have a mango tree in a large self-watering pot, and two years ago noticed that the soil was covered with green. I immediately thought of an algae attack until I examined the growth under a microscope and found hundreds of prothallia from spores of a fan table fern (*Pteris cretica* 'Wilsonii') located about five pots away.

ASEXUAL METHODS OF FERN PROPAGATION

Ferns have three general parts: the leaf or frond, the rhizome or stem, and the roots. Rhizomes either creep along the soil or remain in one place, forming central crowns.

The best time of the year to divide ferns is the spring, before the season's new growth begins. Carefully remove the plant from its pot and shake off the loose soil. If the plant has rambling rhizomes, simply sever with a sharp knife, forming new plants with several fronds and roots. When the crowns are tightly

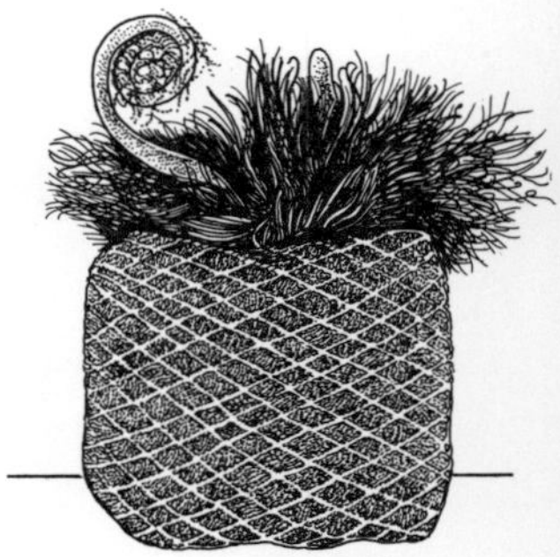

Cutting from a crisped blue fern rooting in a Jiffy-7.

packed together, more care is needed. Use your sharp knife to cut, and try not to disturb the foliage and roots.

Many ferns produce long, bare rhizomes that hang over the edge of the pot. Place a smaller pot next to the mother fern, and lightly cover the tip of a runner with soil, keeping it moist. The runner will produce new fronds and roots, and the new plant may then be severed from the mother.

A few ferns produce plantlets at the tips of the fronds. By pinning the leaf tip on the soil surface of a smaller pot, these plantlets will root, and can then be removed.

CHAPTER 6

Reproducing Cacti and Bromeliads

These plants have been given a chapter of their own because they have germination and propagation methods that differ from the typical houseplant. Neither are too difficult to manage, and you may find them more exciting to raise from seed than the typical flowering plant.

CACTUS

At one time, houseplant growers thought that raising cactus from seed was a waste of time; there were so many on the market at a reasonable cost. Well, times have changed and now it's an increasingly popular hobby and specialization. There are certain differences in the cultural requirements of seedlings, and it might be a year or more before the cactus is established in its own pot, but it's worth the effort.

Seeds are sold as named varieties or as packets of unnamed mixtures. Start with the seed mix; it's the most interesting since you never know what you'll wind up with. Order the seeds to sow in late January or February if you are using a propagating box or individual heating cable. Cacti need a bottom temperature of 21°C (70°F) to insure germination.

The growing medium should be:

2 parts garden or potting soil

1 part shredded peat moss or milled sphagnum moss

1 part sharp sand

The mix must be well sterilized if using garden soil, as cactus seedlings are especially vulnerable to "damping off." In this mix we use some soil rather than man-made products because many of the plants will remain in the original container for at least a year, and the roots do not take to soluble fertilizers.

Use 6-inch clay pots, azalyea-type or 1/2 depth if available, and crock well to insure good drainage. Fill the pots to within 1 inch of the top, tamp the mix, and make sure its smooth and free of lumps. Broadcast the seeds; large seeds should be pushed into the soil by one diameter, with smaller seeds left on top. Now cover the surface with a fine sprinkling of tiny stones—I've found that bird gravel is excellent. These stones help in maintaining moisture and provide the larger seeds with something to push up against, forcing their root deep into the soil.

Water the mix after the seeds are sown. Immerse the pots in water to within 1 inch of the tops, and let them soak for an hour. Now drain well. Cover the pots with a piece of glass or plastic and put them in your propagation box or some source of heat.

The pots can be left in a lighted area, but *never expose them to the direct rays of the sun for at least a year!* In the desert, these tiny plants are protected by larger vegetation (if not, they die).

Germination will begin within three days, though some seeds take up to three months. During this time the mix must always be damp. If it does start to dry, mist the surface early in the morning (allowing water on the seedlings to dry by nightfall) and be sure enough soaks into the soil.

Most of the seeds will produce cotyledons or seed leaves exactly like other plants, but soon the true cactus shape will emerge. Never transplant a seedling; it will have only one root, easily broken, and a good root system doesn't develop until the seed leaves dry up.

Keep the tiny plant thinned allowing 1 inch between each. Transplant to flats when large enough to handle with ease.

The same general instructions apply to growing most succulents from seed.

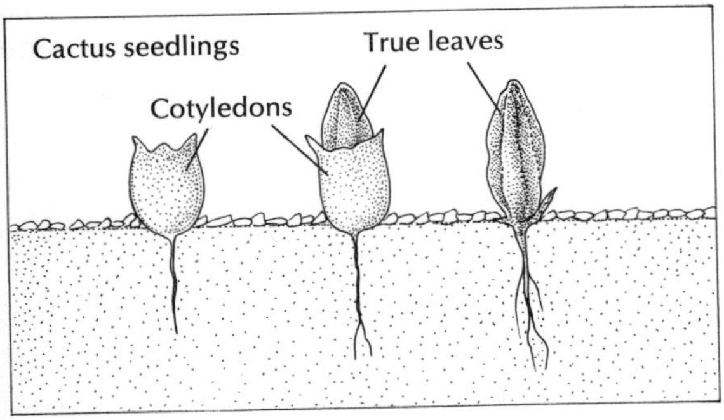

PROPAGATION CACTUS WITH CUTTINGS

Most cacti (and succulents) may be increased with ease by asexual means. The following drawing shows seven types of cacti and succulents and where to cut.

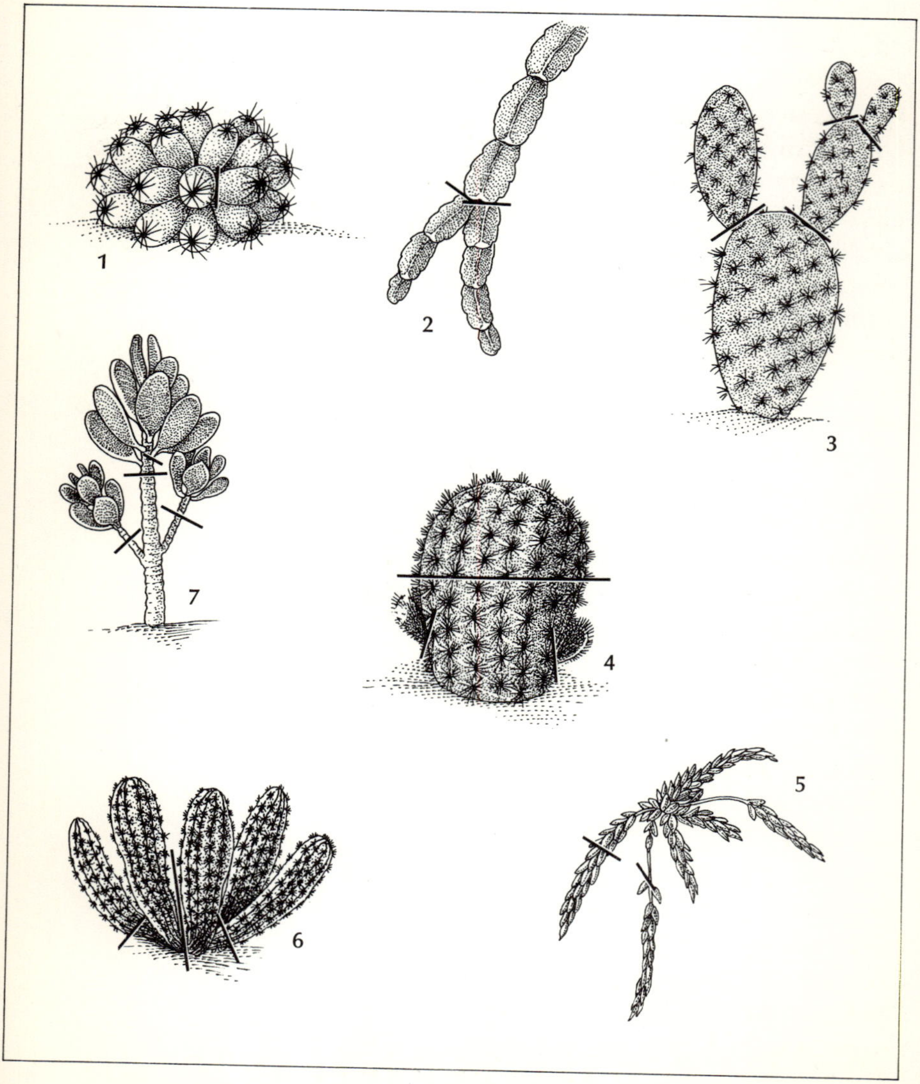

Cuttings should ideally be taken in late spring and early summer, when temperatures are warm and the plants in active growth. Make a clean cut with a sharp knife or razor, and allow the cut to air dry for at least three days, or until a callus or skin forms over the area of the cut.

Use the cactus propagation mix for starting, and carefully push the cuttings into the soil. Don't bury them deeply. If they tend to fall over prop them up with stones. When rooting is well along, transplant to 3-inch pots.

Grafting Cactus

I hesitate to mention grafting because of the abominable things that are done to these plants. Recently, the Orient has exported those cute lollypops of reds and yellows, the Barbie and Kennie dolls of the plant world. You must have seen them, standing in neat rows at variety stores all over the country. The colored tops have no green chlorophyll and cannot manufacture their own food. To remedy this fault, they're grafted to a stock cactus of another species. In fact, one grower now produces kits to make your own. If that's what you want, you'll have to find it on your own, but if you have a problem with a weak-grower or part of a favorite has an incurable disease, the following instructions apply:

Choose cacti of like size and thickness. Cut the top from the desired specimen with a clean, flat cut. Do the same to another cactus, which will act as the roots and stem (this second top could be rerooted rather than thrown away). The top and bottom are gently placed together, the top being held with string weighted with two stones. Now don't move it or let any water in the area for at least a month. The pieces should neatly knit together within this time. The only other requirement is warmth, so summer is the best season to graft.

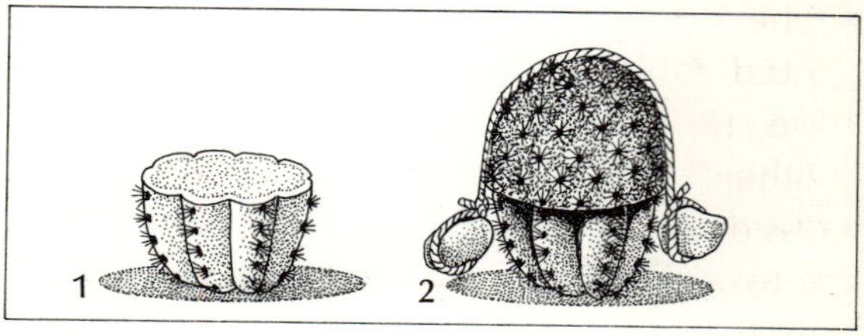

BROMELIADS

Growing bromeliads from seed is not difficult if the seed is reasonably fresh and warm, and if you provide moist tropical conditions. You can buy seed (Thompson & Morgan now list a page of species in their latest catalogue) or produce your own by cross-pollinating flowers.

Since individual flowers last but a day, start early in the morning as flowers open. Remove an anther with a tweezers and allow the pollen to air dry. By mid-morning, the stigma is usually sticky, so apply the pollen with a Q-tip or a camel's hair brush. If fertilization occurs, a seed pod will develop. Some bromeliads are self-sterile, so it's best to use two plants.

The seeds will germinate on a bed of moist and warm sphagnum moss (21°C, 70°F). Cover the container with glass or plastic and place in a lighted area away from sunlight. Thin the seedlings out so that there are 1-inch spaces between them, and transplant when large enough to handle.

Asexual Propagation of Bromeliads

Most bromeliads will bloom only once—the mother plant will then die within a few years of flowering. Before its death, offshoots will appear around the plant's base, each capable of becoming a mature plant.

The offshoots should not be cut off and potted until they are at least 6 inches long and have taken on the characteristics of the mother plant. Root them in a mix of peat moss or sphagnum moss and sand, transplanting to a final home when complete root systems have developed.

The Common Pineapple

Pineapples from the supermarket will make a handsome bromeliad houseplant. Cut off the top growth, taking about 1 inch of fruit. Let the cut air dry for a few days, then place on a bed of moist sand or peat moss. Move to a 5-inch pot with a soil mix of peat, sand, and potting soil, when roots have formed.

Appendix A

Best ways to reproduce 179 different houseplants

(W = root in water)

Archimenes (*Achimenes* spp.) Seeds; scales rubbed from tubers, sown as seeds; shoot cuttings in spring; leaf petiole cutting.

Aeonium (*Aeonium* spp.) Seeds; leaf cuttings; division.

African evergreen (*Syngonium podophyllum*) Stem cuttings. (W)

African milk bush (*Euphorbia tirucalli*) Stem cuttings, but air dry wound first.

African violet (*Saintpaulia* spp.) Seeds; leaf petiole cutting; division. (W)

Aloe (*Aloe* spp.) Suckers; top cuttings.

Aluminum plant (*Pilea cadierei*) Cuttings.

Amaryllis (*Hippeastrum* 'Leopoldii hybrid') Seeds, but will not always breed true; bulblets.

Amazon lily (*Eucharis grandiflora*) Seeds; bulblets.

Anthurium (*Anthurium* spp.) Fresh seed with bottom heat 80°F (26.5°C); top cuttings; division when repotting.

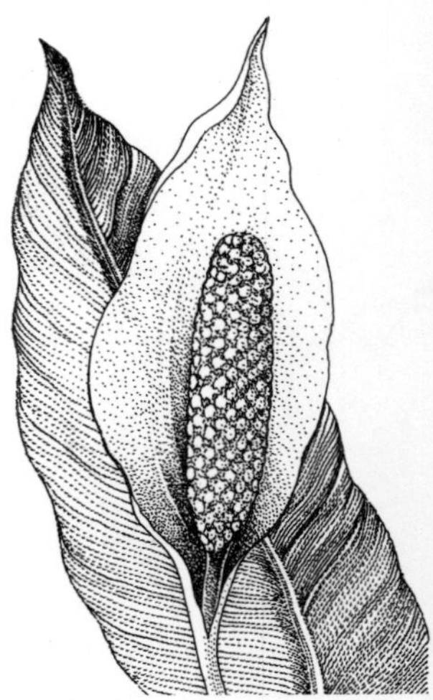

Anthurium
Anthurium andraeanum

Aralia (*Dizygotheca elegantissima*) Seeds; air layer.

Artillery plant (*Pilea microphylla*) Cuttings.

Asparagus fern (*Asparagus* spp.) Seeds; division when repotting.

Aurora borealis plant (*Kalanchoe fedtschenkoi marginata*) Stem cuttings, but water sparingly.

Avocado (*Persea americana*) Seeds; stem cuttings in spring. (W)

Azalea (*Rhododendron* x hybridia) Top cuttings from new growth.

Banana (*Musa nana*) Seeds, but may take a year to sprout; suckers.

Begonia (*Begonia* spp.):

a) Fibrous-rooted or bedding begonias by seeds; stem cuttings; division. (W)
b) Cane-stemmed begonias by stem cuttings.
c) Rhizomatious begonias by division of rhizome including an "eye" in each section; tip cuttings.
d) Rex begonias by seeds; tip cuttings; leaf section cuttings.
e) Tuberous-rooted begonias by seeds; division of tuber with "eye" in each section.

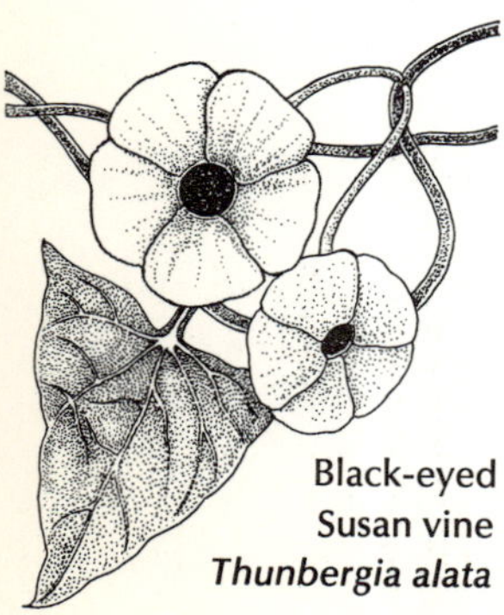

Black-eyed Susan vine *Thunbergia alata*

Bird of paradise (*Strelitzia reginae*) Seeds; division of new shoots.

Black-eyed susan vine (*Thunbergia alata*) Seeds; cuttings.

Bloodleaf (*Iresine herbstii*) Cuttings. (W)

Blood lily (*Haemanthus katherinae*) Bulblets.

Blue marguerite (*Felicia amelloides*) Seeds; cuttings; division.

Bottlebush (*Callistemon lanceolatus*) Seeds; stem cuttings of ripened wood.

Bougainvillea (*Bougainvillea* spp.) Seeds; stem cuttings from ripened wood.

Brazilian firecracker (*Manettia inflata*) Stem cuttings.

Browallia (*Browallia speciosa*) Seeds; cuttings.

Bougainvillea
Bougainvillea glabra

Caladium (*Caladium* spp.) Divide tubers in spring with bottom heat of 70°F (21°C).

Calathea (*Calathea* spp.) Division in spring.

Calico plant (*Alternanthera* spp.) Cuttings in spring; division.

Camellia (*Camellia* spp.) Tip cuttings in fall with rooting hormone.

Campanula (*Campanula isophylla*) Cuttings.

Carrion flower (*Stapelia* spp.) Seeds; cuttings; division.

Cast-iron plant (*Aspidistra elatior*) Division.

Cattleya orchid (*Cattleya* spp.) Division with at least four pseudobulbs to a plant.

Century plant (*Agave* spp.) Seeds; suckers.

Cherry tomato (*Lycopersicon esculentum* 'Tiny Tim') Seeds from your own fruit.

Chinese evergreen (*Aglaonema* spp.) Seeds; division of mature plant; joint cuttings.

Christmas cactus (*Zygocactus* spp.) Leaf joint.

Christmas kalanchoe (*Kalanchoe blossfeldiana*) Seeds; stem cuttings.

Cigar plant (*Cuphea platycentra*) Seeds, but need light to germinate; cuttings.

Cissus (*Cissus* spp.) Cuttings with a piece of the older branch. (W)

Citrus (*Citrus* spp.) Seeds, but not always true to type; cuttings in spring.

Clerodendrum (*Clerodendrum* spp.) Tip cuttings.

Cobra lily (*Darlingtonia californica*) Seeds; division of mature plants.

Coffee (*Coffea arabica*) Seeds; firm cuttings with bottom heat of 80°F (26.5°C).

Coleus (*Coleus* spp.) Seeds; cuttings. (W)

Columnea (*Columnia* spp.) Tip cuttings; division; single leaves will eventually root, but it takes forever.

Coral bead plant (*Nertera granadensis*) Seeds; division.

Coral vine (*Antigonon leptopus*) Seeds; cuttings.

Crotons (*Codiaeum var. pictum*) Seeds, but never true to type; tip cuttings.

Crown of thorns (*Euphorbia splendens*) Stem cuttings, but allow wound to heal first.

Cyclamen (*Cyclamen* spp.) Seeds sown at 65°F (18°C) and when germinated, grown at 55°F (13°C); tuber cuttings.

Cyclamen
Cyclamen neapolitanum

Devil's ivy (*Scindapsus* spp.) Tip cuttings.

Dracaena (*Dracaena* spp.) Cuttings; stem sections; air layer. (W)

Drunkard's dream (*Hatiora salicornioides*) Stem cuttings of joints.

Dumbcanes (*Dieffenbachia* spp.) Stem sections; suckers; air layer. (W)

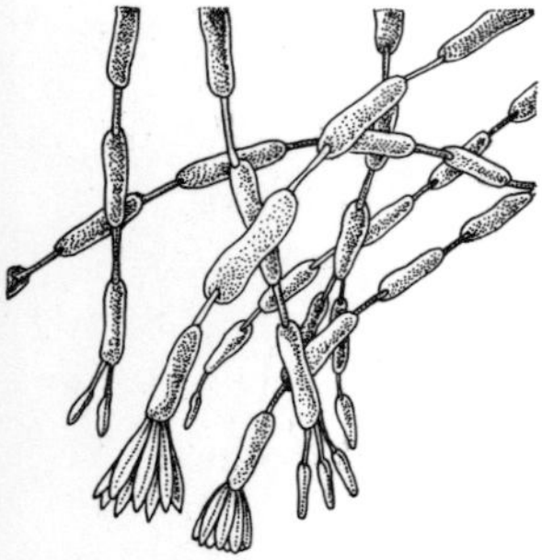

Drunkard's dream
Hatiora salicornioides

Echeveria (*Echeveria* spp.) Cuttings; leaf cuttings; offshoots.

Egyptian star cluster (*Pentas lanceolata*) Seeds; softwood cuttings in spring.

English daisy (*Bellis perennis florepleno*) Seeds; division.

English nasturtium (*Tropaeolum majus florepleno*) Seeds; cuttings.
Exacum (*Exacum affine*) Seeds; cuttings.

Fern, boston (*Nephrolepis exaltata*) Rooting buds on runners; division.
Fern, crisped blue (*Polypodium aureum* 'Mandaianum') Break off a large, hairy "foot" and pot separately.
Fern, mother (*Asplenium bulbiferum*) Tiny plantlets develop on frond tops, to be potted.
Fern, rabbit's foot (*Davallia canariensis*) Break off a "foot" and repot.
Fern, staghorn (*Platycerium* spp.) Tiny "pups" at root tips may be potted.
Fig, creeping (*Ficus pumila*) Stem cuttings.
Fig, fiddleleaf (*Ficus lyrata*) Air layer.
Fig, fruit (*Ficus carica*) Hardwood cuttings.
Fire cracker plant (*Gesneria cuneifolia*) Seeds; leaf cuttings in summer; stem cuttings.
Flamebush (*Calliandra emarginata*) Seeds; new stem cuttings.
Flame violets (*Episcia* spp.) Seeds; cuttings; rosettes from runners.
Flamingo flower (*Jacobinia carnea*) Stem cuttings.
Flowering maple (*Abutilon* x hybridia) Seeds; stem cuttings from new growth in spring.
Freesia (*Freesia* x hybridia) Cormlets.
Forest lily (*Veltheimia viridifolia*) Bulblets; single, whole leaf, stuck upright in medium will produce bulblets around base.
Fuchsia (*Fuchsia* x hybridia) Seeds; stem cuttings in spring.

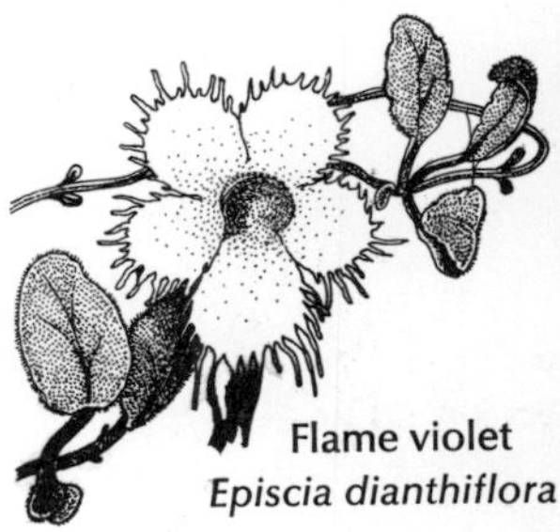

Flame violet
Episcia dianthiflora

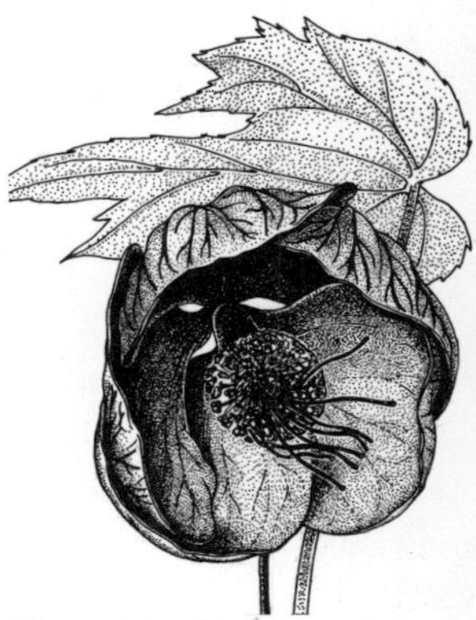

Flowering maple
Abutilon x hybridia

Gardenia (*Gardenia* spp.) Half-hard cuttings from new growth with bottom heat of 75°F (24°C).

Gas plant (*Dictamnus albus*) Seeds; root cuttings; division.

Ganzania (*Gazania rigens*) Seeds; cuttings; division.

Geranium (*Pelargonium* spp.) Stem cuttings.

German ivy (*Senecio mikanioides*) Stem cuttings. (W)

Gill-over-the-ground (*Glecoma hederacea*) Cuttings.

Glory lily (*Gloriosa rothschildiana*) Seeds; cut tuber including an "eye," dusting cut end with powdered charcoal.

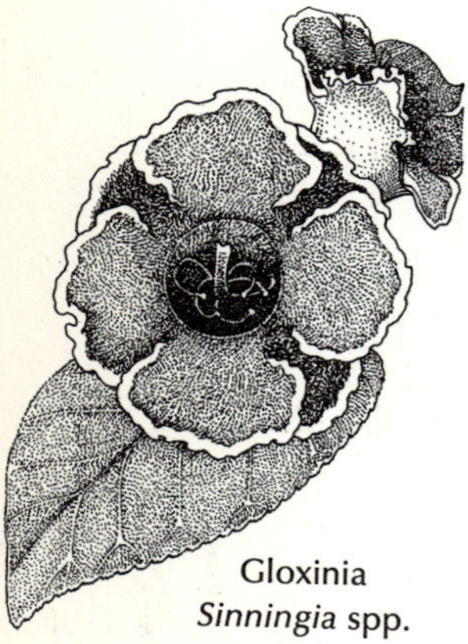

Gloxinia
Sinningia spp.

Gloxinia (*Sinningia* spp.) Seeds; sprouts; leaf cuttings; division of tuber with "eye."

Golden trumpet (*Allamanda cathartica hendersonii*) Softwood cuttings in spring.

Good-luck plant (*Kalanchoe* 'daigremontiana x tubiflora') Tiny plantlets fall to earth and grow.

Green marble vine (*Senecio herreianus*) Division; cuttings.

Guinea goldvine (*Hibbertia volubilis*) Stem cuttings.

Hardy chinese orchid (*Bletilla striata*) Division during dormant period.

Heliotrope (*Heliotropium arborescens*) Seeds; stem cuttings.

Hibiscus (*Hibiscus rosa-sinensis*) Stem cuttings of firm young shoots in spring with bottom heat of 70°F (21°C).

Hoya (*Hoya* spp.) Stem cuttings of previous year's growth in spring with bottom heat of 75°F (23°C).

Hydrangea (*Hydrangea macrophylla*)
Cuttings of young shoots with at least three nodes in spring.

Ivy (*Hedera helix*) Cuttings. (W)
Impatiens (*Impatiens* spp.) Seeds; cuttings. (W)
Inch plant, also called Wandering Jew (*Tradescantia* spp.) Cuttings. (W)

Jade plant (*Crassula* spp.) Stem cuttings; leaf cuttings. (W)
Japanese aralia (*Fatsia japonica*) Seeds; stem cuttings; air layer.
Jasmine (*Trachelospermum jasminoides*) Stem cuttings.
Jerusalem cherry (*Solanum pseudo-capsicum*) Seeds; stem cuttings.
Jungle flame (*Ixora macrothyrsa*) Firm stem cuttings in spring at bottom heat of 85°F (29°C).

Kenilworth ivy
Cymbalaria muralis

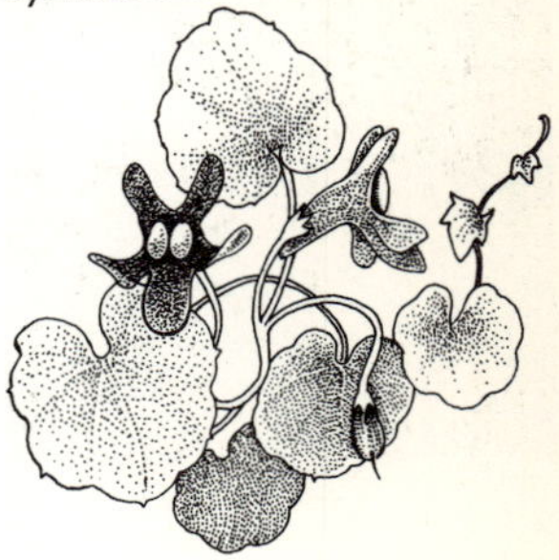

Kafir lily (*Clivia miniata* 'Grandiflora') Seeds; division when repotting.
Kenilworth ivy (*Cymbalaria muralis*) Seeds; cuttings. (W)

Lantana (*Lantana camara*) Seeds; softwood cuttings.
Leopard plant (*Ligularia tussilaginea*) Division when repotting.
Lily (*Lilium* spp.) Seeds; bulblets; bulbils that appear at leaf axils; individual scales from mother bulb laid flat on sphagnum moss.
Lily-of-the-Nile (*Agapanthus africanus*) Seeds; division in spring when repotting.

Lipstick vine (*Aeschynanthus lobbianus*) Stem cuttings.

Lobster claw (*Heliconia humilis*) Seeds; division.

Miracle leaf (*Kalanchoe pinnata*) Plantlets formed in leaf notches may be potted.

Monkey flower (*Mimulus aurantiacus*) Seeds; stem cuttings.

Monstera (*Monstera deliciosa*) Seeds; stem sections; stem cuttings. (W)

Moon Flower (*Calonyction aculeatum*) Seeds, but notch the seed coat to hasten germination; stem cuttings but chancey.

Moses-in-the cradle (*Rhoeo spathacea*) Seeds; suckers.

Nerve plant
Fittonia vershaffelitti

Natal plum (*Carissa grandiflora*) Seeds; stem cuttings.

Nerve plant (*Fittonia vershaffeltti*) Stem cuttings with bottom heat 75°F (23°C).

Night-blooming cereus (*Selenicereus* spp.) Joint or tip cuttings.

Norfolk Island pine (*Araucaria heterophylla*) Seeds; 6-inch tip cuttings from terminal shoots.

Norse fire plant (*Columnea* x 'Stavanger') Division; stem cuttings.

Oleander (*Nerium oleander*) Stem cuttings. (W)

Orchid cactus (*Epiphyllum* spp.) Cuttings of mature shoots in spring.

Oxalis (*Oxalis* spp.) Seeds; offsets when repotting.

Palms (all varieties) Fresh seeds with bottom heat of 70°F (21°C); suckers.

Panda plant (*Kalanchoe tomentosa*) Stem cuttings; individual leaves on moist sand.

Parachute plant (*Ceropegia sandersonii*) Joint cuttings.

Passion flower (*Passiflora* spp.) Seeds; stem cuttings.

Peperomia (*Peperomia* spp.) Soft wood cuttings; division.

Petunia (*Petunia* x hybridia) Seeds; cuttings.

Philodendron (*Philodendron* spp.) Seeds; tip cuttings; air layer; stem sections with several leaves. (W)

Piggy-back plant (*Tolmiea menziesii*) Leaf petiole cuttings. (W)

Pineapple lily (*Eucomis zambesiaca*) Bulblets.

Pitcher plants (*Sarracenia* spp.) Seeds; division in early spring.

Plover eggs (*Adromischus festivus*) Stem cuttings; individual leaves on moist sand.

Plumbago (*Plumbago capensis*) Tip cuttings; root cuttings; division when repotting.

Pocketbook plant (*Calceolaria* spp.) Seeds; stem cuttings.

Poinsettia (*Euphorbia pulcherrima*) Soft or hardwood cuttings allowing the wound to heal.

Prayer plant (*Maranta leuconeura*) Division when repotting.

Pitcher plant
Sarracenia flava

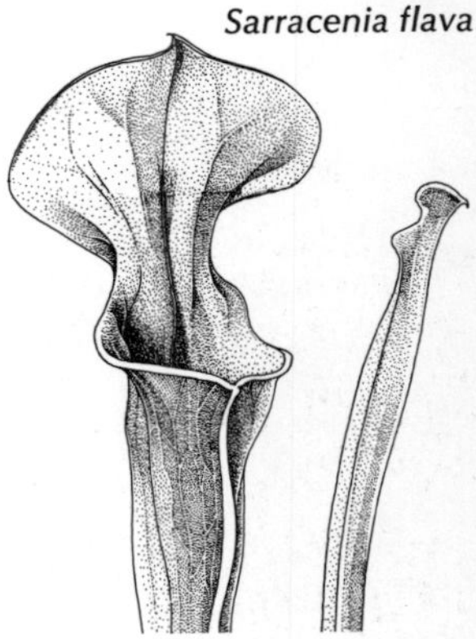

Quilted taffeta plant (*Hoffmannia roezlii*) Stem cuttings.

Resurrection plant (*Selaginellia lepidophylla*) Pin a frond tip to damp sand or peat moss and new plant will form.

Rochea coccinea Stem cuttings, air dryed for a few days in spring or summer.

Rouge plant (*Rivina humilis*) Seeds; stem cuttings.

Rubber plant (*Ficus elastica*) Seeds; air layer.

Sacred Bo-tree (*Ficus religiosa*) Seeds; air layer.

Scarborough lily (*Vallota speciosa*) Seeds; bulblets.

Screw pine (*Pandanus* spp.) Suckers.

Sea onion (*Ornithogalum caudatum*) Bulblets.

Shrimp plant (*Beloperone guttata*) Stem cuttings.

Snake plant (*Sansevieria trifasciata*) Division; cross-sections of leaves.

Spanish shawl (*Schizocentron elegans*) Stem cuttings.

Spider lily (*Hymenocallis* spp.) Seeds; Bulblets.

Spider plants (*Chlorophytum comosum*) Root plantlets at leaf tips; division when repotting.

Strawberry geranium (*Saxifraga sarmentosa*) Fasten plantlets at tip of runners on moist soil.

Streptocarpus (*Streptocarpus* spp.) Seeds; leaf cuttings; leaf sections.

Swedish ivy (*Plectranthus australis*) Stem cuttings. (W)

Sweet flag (*Acorus gramineus variegatus*) Division of rootstock. (W)

Sweet-potato vine (*Ipomoea batatas*) Stem cuttings; division of tubers but include an "eye."

Sundews (*Drosera* spp.) Seeds; division.

Streptocarpus
Streptocarpus rexii

Tahitian bridal veil (*Gibasis geniculata*) Stem cuttings. (W)

Temple Bells (*Smithiantha* spp.) Seeds; rhizome cuttings; leaf cuttings without the petiole.

Thanksgiving cactus (*Zygocactus truncatus*) Leaf joints.

Ti plant (*Cordyline terminalis*) Air layer; stem sections; root sections. (W)

Umbrella plant (*Cyperus alternifolius*) Seeds; division; tip cuttings. (W)

Velvet Plant (*Gynura auranticica*) Stem cuttings. (W)

Venus fly-trap (*Dionaea muscipula*) Seeds; division; leaf cuttings.

Walking iris (*Neomarica gracilis*) Division; root plantlets at end of leaves.

Wandering Jew (*Zebrina* spp.) Stem cuttings. (W)

Watsonia (*Watsonia humilis maculata*) Bulblets.

Wood rose (*Ipomoea tuberosa*) Seeds, but notch seed coat before sowing; Stem cuttings.

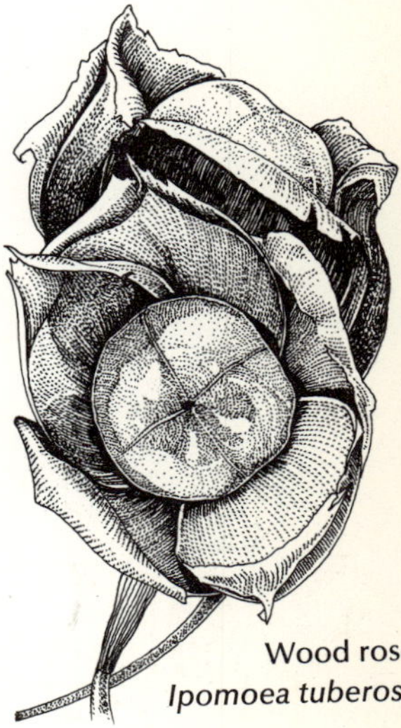

Wood rose
Ipomoea tuberosa

Zebra plant (*Aphelandra* spp.) Seeds; Stem cuttings in spring at 70°F (21°C).

Zebra plant (*Calathea zebrina*) Division when repotting.

Appendix B

Some seeds for houseplants that can be started any time of the year

(F = flowers • L = foilage)

Achimenes spp. (F)
Aeonium spp. (L)
African daisy (*Gerbera jamesonii*) (F)
African violet (*Saintpaulia* spp.) (F)
Amaryllis (*Hippeastrum* spp.) (F)
Amazon lily (*Eucharis grandiflora*) (F)
Anthurium spp. (F,L)
Aralia (*Dizygotheca elegantissima*) (L)
Asparagus fern (*Asparagus* spp.) (L)
Avocado (*Persea americana*) (L)

Balsam (*Impatiens balsamina*) (F)
Banana (*Musa nana*) (L)
Begonia (*Begonia* spp.) (F,L)
Bird of paradise (*Strelitzia reginae*) (F)
Black-eyed susan vine (*Thunbergia alata*) (F)
Blue marguerite (*Felicia amelloides*) (F)
Bottlebush (*Callistemon lanceolatus*) (F)
Bougainvillea spp. (F)
Browallia spp. (F)
Butterfly pea (*Clitoria ternatea*) (F)

Blue marguerite
Felicia amelloides

Coral bead plant
Nertera granadensis

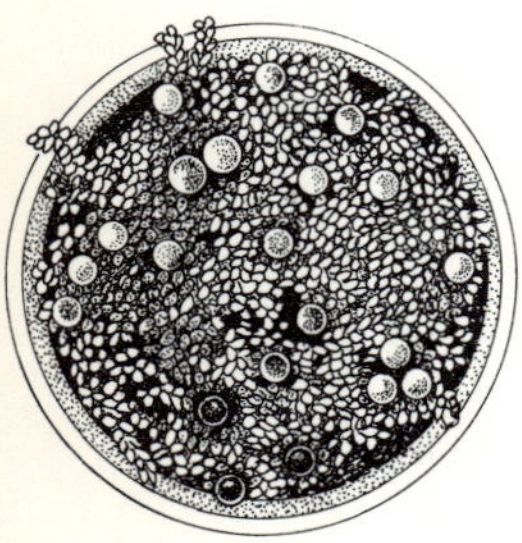

Cacti, most varieties (F,L)
Calendula (*Calendula officinalis*) (F)
Cardinal climber (*Quamoclit sloteri*) (F,L)
Century plant (*Agave* spp.) (L)
Cherry tomato (*Lycopersicon esculentum*) (F)
Chinese evergreen (*Aglaonema* spp.) (L)
Christmas kalanchoe (*Kalanchoe blossfeldiana*) (F,L)
Christmas pepper (*Capsicum annum*) (F,L)
Cigar plant (*Cuphea platycentra*) (F)
Cobra lily (*Darlingtonia californica*) (F,L)
Coffee (*Coffea arabica*) (L)
Coleus (*Coleus* spp.) (L)
Coral bead plant (*Nertera granadensis*) (F,L)
Coral vine (*Antigonon leptopus*) (F,L)
Crotons (*Codiaeum* var. *pictum*) (L)
Cyclamen (*Cyclamen* spp.) (F,L)
Cypress vine (*Quamoclit pennata*) (F,L)

English daisy (*Bellis perennis florepleno*) (F)
English nasturtium (*Tropaeolum majus florepleno*) (F,L)
Egyptian star cluster (*Pentas lanceolata*) (F)
Exacum (*Exacum affine*) (F)

Ferns, most varieties, but *spores* not *seeds* (L)
Firecracker plant (*Gesneria cuneifolia*) (F)
Flag of Spain (*Quamoclit lobata*) (F,L)
Flamebush (*Calliandra emarginata*) (F,L)
Flame violets (*Episcia* spp.) (F,L)
Flowering maple (*Abutilon* x hybridia) (F,L)
Fuchsia (*Fuchsia* x hybridia) (F)

Gas plant (*Dictamnus albus*) (F,L)
Gazania (*Gazania rigens*) (F)
Glory lily (*Gloriosa rothschildiana*) (F)

Glory pea (*Clianthus formosus*) (F)
Gloxinia (*Sinningia* spp.) (F)

Heliotrope (*Heliotropium arborescens*) (F)

Ice plant (*Mesembryanthemum crystallinum*) (F,L)
Impatiens (*Impatiens* spp.) (F)

Japanese aralia (*Fatsia japonica*) (L)
Jasmine (*Jasminum* spp.) (F)
Jerusalem cherry (*Solanum pseudo-capsicum*) (F)

Kafir lily (*Clivia miniata* 'Grandiflora') (F,L)
Kenilworth ivy (*Cymbalaria muralis*) (F,L)

Lantana (*Lantana camara*) (F)
Lily (*Lilium* spp.) (F)
Lily-of-the-Nile (*Agapanthus africanus*) (F,L)
Living stones (*Lithops* spp.) (F,L)
Lobster claw (*Heliconia humilis*) (F,L)

Mandevilla suaveolens (F,L)
Marble vine (*Bryonopsis laciniosa*) (F,L)
Mimosa (*Acacia* spp.) (F,L)
Monkey flower (*Mimulus aurantiacus*) (F)
Monstera (*Monstera deliciosa*) (F,L)
Moon flower (*Calonyction aculeatum*) (F)
Morning glories (*Ipomoea* spp.) (F,L)
Moses-in-the-cradle (*Rhoeo spathacea*) (L)

Nasturtium (*Tropaeolum* spp.) (F,L)
Natal plum (*Carissa grandiflora*) (F,L)
Norfolk Island pine (*Araucaria heterophylla*) (L)

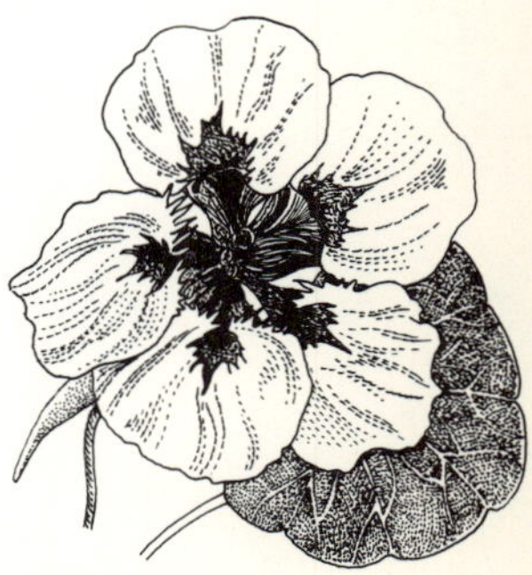

Nasturtium
Tropaeolum spp.

Oleander (*Nerium oleander*) (F)
Oxalis (*Oxalis* spp.) (F)

Palms, most varieties
Passion flower (*Passiflora* spp.) (F)
Petunia (*Petunia* x hybridia) (F)
Philodendron (*Philodendron* spp.) (L)
Pocketbook plant (*Calceolaria* spp.) (F)

Pocketbook plant
Calceolaria spp.

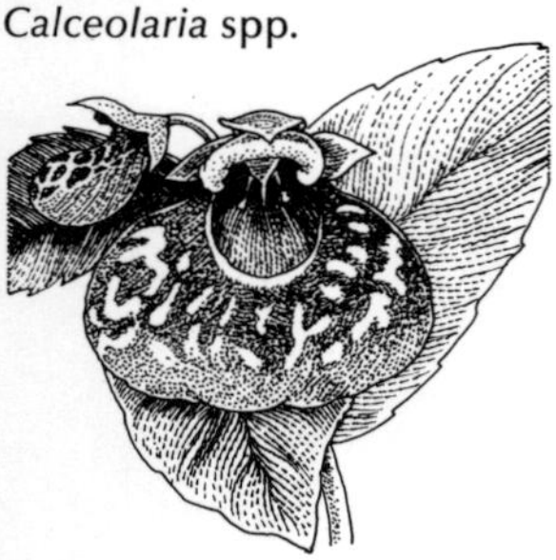

Rouge plant (*Rivina humilis*) (F,L)
Rubber plant (*Ficus elastica*) (L)

Sacred Bo-tree (*Ficus religiosa*) (L)
Scarborough lily (*Vallota speciosa*) (F)
Spider lily (*Hymenocallis* spp.) (F)
Streptocarpus (*Streptocarpus* spp.) (F)
Sundews (*Drosera* spp.) (L)

Temple Bells (*Smithiantha* spp.) (F)
Torenia fournieri (F)
Transvaal daisy (*Gerbera jamesonii*) (F)

Umbrella plant (*Cyperus alternifolius*) (L)

Venus fly-trap (*Dionaea muscipula*) (L)

Wood rose (*Ipomoea tuberosa*) (F)

Zebra plant (*Aphelandra* spp.) (F,L)

Appendix C

Month-by-month Projects

January

The Druids called it the "Black Month," but there's still plenty to do.

Plant seeds of tuberous rooted begonias for July bloom, achimenes for fall bloom, and African violets, begonias, blue Marguerite, browallia, gloxinia, impatiens.

Divide and repot African violets, begonias, clerodendrum.

Make cuttings of clerodendrum, crassula, gardenia, lantana.

February

Weather is bleak, but remember days are now getting longer.

Plant seeds of African daisy for fall bloom, and African violets, azalea, begonia, cactus, freesia, kalanchoe, spathiphyllum.

Divide and repot African violets, begonias, flowering maple, lily-of-the-Nile, orchids.

Plant bulbs of amaryllis, calla lilies, caladium, gloxinia.

Make cuttings of coleus, crassula, gardenia, lantana, shrimp plant.

March

The worst of winter is over.

Plant seeds of anthurium, begonia, cactus, clivia, flowering maple, temple bells.

Divide and repot allamanda, cactus, columnea, orchids.

Plant bulbs of achimenes, tuberous begonias, caladium.

Make cuttings of coleus, columnea, crassula, hoya, jasmine, lantana, passion flower.

April

Time to provide shade for those plants needing protection from the sun.

Plant seeds of most houseplants, ornamental peppers, streptocarpus.

Divide and repot all plants in need, before growth really takes off.

Plant bulbs of Amazon lilies, gloxinias, Peruvian daffodils, Scarborough lilies.

Make cuttings of bougainvillea, columnea, crassula, croton, Marguerites.

NOTE: Put cyclamens in a cool, bright area and remember to keep them growing; prune and repot Christmas Poinsettias, then take cuttings of new growth for winter bloom.

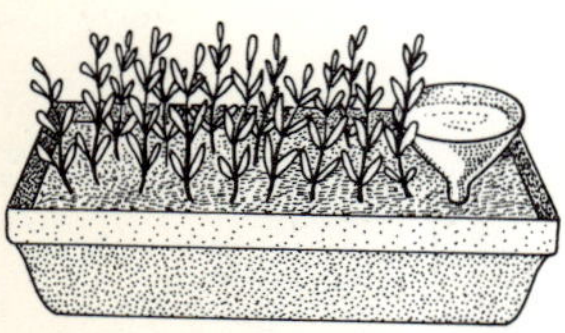

Watering funnel

May

Plan to take houseplants outdoors for the summer, after the last frost in your area.

Plant seeds of most houseplants.

Divide and repot anything that needs it, and watch your new plants grown from seeds for signs of outgrowing present pots.

Make cuttings of aphelandra, African violets, and most everything else.

NOTE: Don't forget to fertilize seedlings grown in an artificial medium.

June

Summer is here, so watch heat and light on your new plants.

Plant seeds of calceolaria, cineraria, cyclamen, gloxinias, impatiens.

Divide and repot anything that needs it.

Make cuttings of most foliage plants.

July

Same comments as June.

Plant seeds of calendula, coleus, impatiens, kalanchoe.

August

Fall catalogues should be arriving; start making selections on flowering bulbs for winter.

Plant seeds of bellis, calendula, cineraria, nasturtium, schizanthus.

Divide and repot African violets, begonias, orchids.

Plant bulbs of freesia, lachenalia, oxalis.

Make cuttings of azalea, camellia, coleus, crassula, fuchsia, impatiens.

September

Time to take cuttings of garden plants like edging begonias and geraniums for winter bloom.

Plant bulbs for winter forcing.

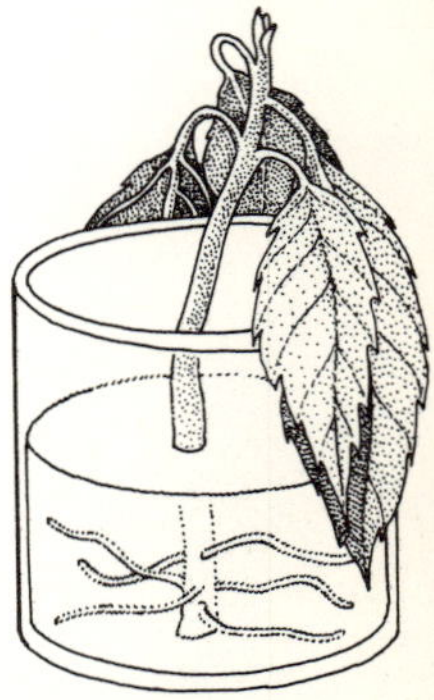

Cutting in water

October

In most parts of the country, the killing frost is sure to arrive. Take cuttings of the last of the garden plants: begonias, bloodleaf, fuchsia, geraniums, heliotrope, lemon verbena, periwinkle.

NOTE: Your poinsettia cuttings from last spring must now be shielded from any artificial light to insure bloom for Christmas.

November

Remember to protect cuttings and seedlings from the coming cold.

Plant seeds of achimenes and African violets.

Plant bulbs of crocus, daffodil, hyacinth, oxalis, tulip.

December

The year's almost over and spring catalogues will soon be arriving.

Plant seeds of African violet, begonia, lobelia.

Divide and repot African violet, orchids.

Plant bulbs of amaryllis and start tubers of gloriosa lily.

Appendix D

Sources of seeds and equipment

W. Atlee Burpee Co.
Warminster, Pennsylvania 18974
Clinton, Iowa 52732
Riverside, California 92502
Seeds; Equipment for germination
Catalogue: Free

J.L. Hudson, Seedsman
P.O. Box 1058
Redwood City, California 94064
Seeds, all varieties
Catalogue: $.50

J. W. Jung Seed Co.
Randolph, Wisconsin 53956
Some houseplant seeds
Catalogue: Free

George W. Park Seed Co., Inc.
Greenwood, South Carolina 29647
Many seeds; Equipment for germination
Catalogue: Free

Plant labels

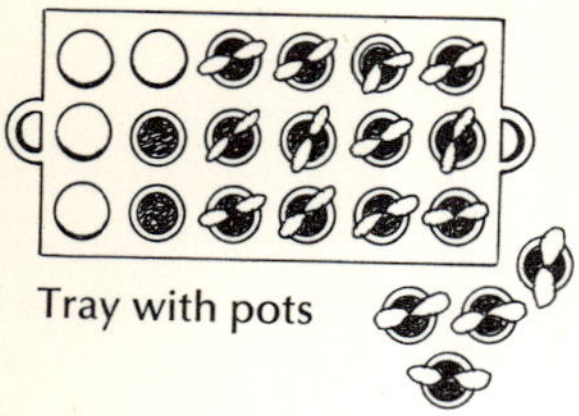

Tray with pots

Thompson & Morgan
P.O. Box 24
401 Kennedy Blvd.
Somerdale, New Jersey 08083
Many kinds of houseplant seeds
Catalogue: Free

Equipment only

House Plant Corner
Box 5000
Cambridge, Maryland 21613
All types of equipment
Catalogue: Free

Mellinger's Inc.
2310 West South Ridge Road
North Lima, Ohio 44452
All types of equipment
Catalogue: Free

Walter F. Nicke
Box 71
Hudson, New York 12534
All types of equipment
Catalogue: $.25

Propagating pot

Appendix E

Societies and magazines for special plant groups

The following plant societies all publish specialty magazines and newsletters, often with much information on seeds and propagation:

The American Begonia Society
14050 Ramona Drive
Whittier, California 90605
Begonian Magazine issued monthly
$4.00 per year.

American Fern Society
Department of Botany
University of Tennessee
Knoxville, Tennessee 37916
Journal on ferns issued quarterly; maintains spore exchange for members.
$5.00 per year

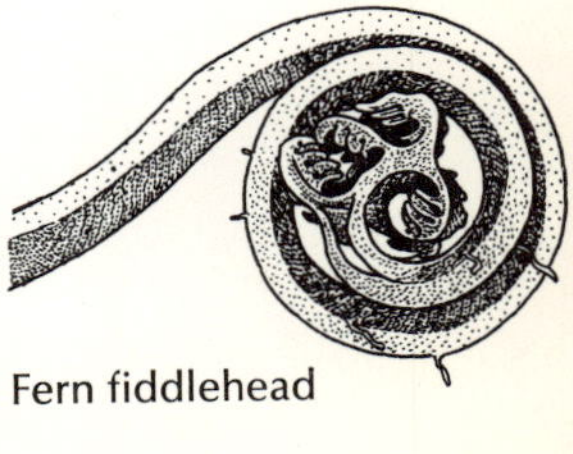

Fern fiddlehead

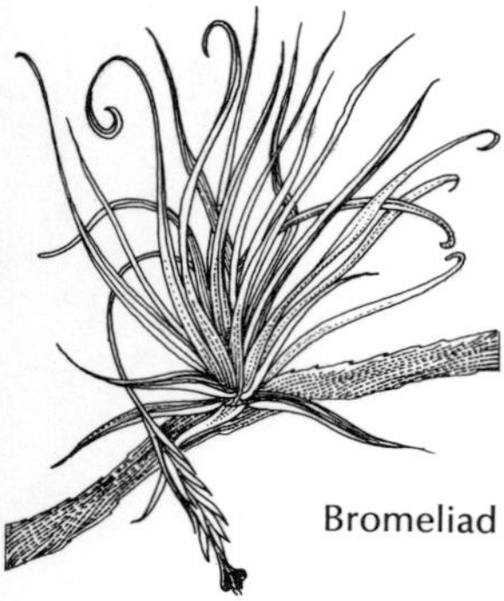

Bromeliad

The Bromeliad Society
P. O. Box 3279
Santa Monica, California 90403
Bromeliad Journal issued bi-monthly
$7.50 per year.

The Cactus and Succulent Society
Box 167
Reseda, California 91335
Journal issued bi-monthly
$7.50 per year.

The Indoor Light Gardening Society
128 West 58th Street
New York, New York 10019
Light Garden issued bi-monthly
$5.00 per year.

The International Fern Society
2423 Burritt Avenue
Redondo Beach, California 90278
Newsletter issued monthly
$3.50 per year

Massachusetts Horticultural Society
300 Massachusetts Avenue
Boston, Massachusetts 02115
Horticulture issued monthly
$8.00 per year

Plants Alive issued monthly
319 NE 45th Street
Seattle, Washington 98105
$9.00 per year

The American Horticultural Society
7931 East Boulevard Drive
Alexandria, Virginia 22308
This Society has a great seed sharing offer to its members every year and publishes *American Horticulturist* six times a year.
$15.00 per year

Appendix F

Temperature Conversion Chart

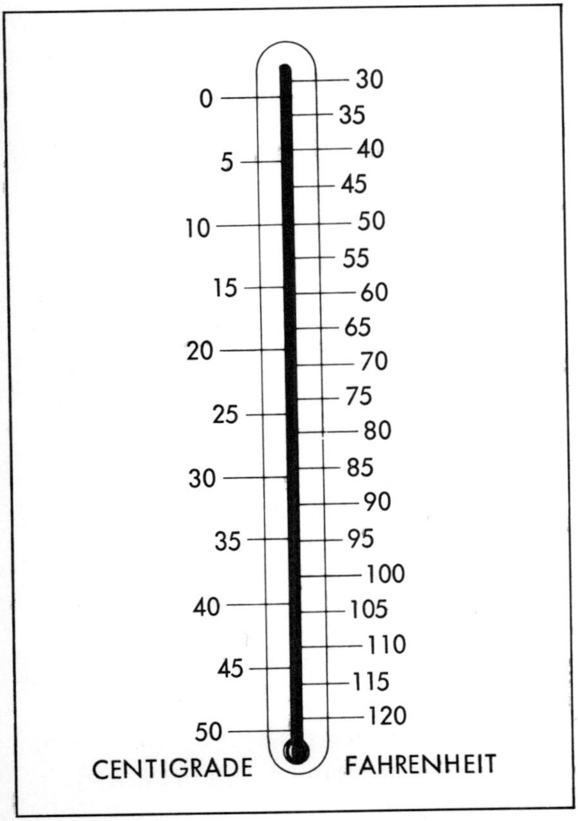

Bibliography

Bailey, L.H., and Bailey, E.Z.
Hortus Second
New York: Macmillan Publishing Co., Inc., 1974

Britton, N.L., and Rose, J.N.
The Cactaceae. 4 vols. 1937
Reprint (4 vols. in 2)
New York: Dover Publications, 1963

Cobb, Boughton
A Field Guide to the Ferns
Boston: Houghton Mifflin Company, 1956

Graf, Alfred Byrd
Exotica 6
New Jersey: Roehr's Company, 1973

Graf, A.B.
Exotic Plant Manual
New Jersey: Roehr's Company, 1970

Hartman, H.T., and Kester, D.E.
Plant Propagation
New Jersey: Prentice-Hall, Inc. 1975

Lamb, Edgar and Brian
The Pocket Encyclopedia of Cacti and Succulents
New York: The Macmillan Company, 1970

Loewer, H.P.
Bringing the Outdoors In
New York: Walker and Company, 1974

Wells, James S.
Plant Propagation Practices
New York: The Macmillan Company, 1971

Wilson, Robert Gardner and Catherine
Bromeliads in Cultivation
Coconut Grove, Florida:
Hurricane House Publishers, Inc. 1964

Wyman, Donald
Wyman's Gardening Encyclopedia
New York: The Macmillan Company, 1971